J.G. Taylor (Ed.)

# NEURAL NETWORK APPLICATIONS

Proceedings of the Second British Neural Network
Society Meeting (NCM91), London, October 1991

Springer-Verlag
London Berlin Heidelberg New York
Paris Tokyo Hong Kong
Barcelona Budapest

J. G. Taylor, BA, BSc, MA, PhD, FInstP
Director, Centre for Neural Networks,
Department of Mathematics, King's College,
Strand, London WC2R 2LS, UK

*Series Editors*

J. G. Taylor, BA, BSc, MA, PhD, FInstP
Director, Centre for Neural Networks,
Department of Mathematics, King's College,
Strand, London WC2R 2LS, UK

C. L. T. Mannion, BSc, PhD, MInstP
Department of Electrical Engineering, University of Surrey, Guildford,
Surrey, GU2 5XH, UK

ISBN-13:978-3-540-19772-0     e-ISBN-13:978-1-4471-2003-2
DOI: 10.1007/978-1-4471-2003-2

British Library Cataloguing in Publication Data
Neural Network Applications: Proceedings of the Second British Neural Network
Society Meeting (NCM '91), London, October 1991. – (Perspectives in Neural
Computing Series)
I. Taylor, John   II. Series
006.3
ISBN-13:978-3-540-19772-0

Library of Congress Data available

Typesetting: Camera ready by author
34/3830-543210 Printed on acid-free paper

# PREFACE

This volume contains 12 papers on the applications of neural networks. The papers were given at the one-day neural computing meeting, NCM91, on 1 October, 1991 at King's College London, which was sponsored by the Centre for Neural Networks, King's College London, and the British Neural Nets Society, and was also part of the DEANNA (Database for European Artificial Neural Network Activity) ESPRIT programme. We would like to thank these bodies for their support.

The papers reflect the spectrum of applications that are presently being attempted in neural networks. They cover medical diagnosis, industrial and control areas, as well as more general discussions on net learning and knowledge representation. The breadth and depth of coverage is a sign of the health of the subject, as well as indicating the importance of neural network developments in industry. It is hoped this volume will be of value to those attempting to apply neural networks, as well as those entering the field.

January 1992                                                                 J.G. Taylor

# CONTENTS

# CONTRIBUTORS

Anthony, D.M.
19 Winifred Avenue, Earlsdon, Coventry CV5 6JT, UK

Austin, J.
Advanced Computer Architecture Group, Department of Computer Science,
University of York, Heslington, York YO1 5DD, UK

Biggs, N.L.
Department of Mathematics, London School of Economics, Houghton Street,
London WC2A 2AE, UK

Bishop, C.
AEA Technology, Instrumentation Division, Harwell Laboratory, Oxfordshire OX11 0RA, UK

Boyce, J.F.
Department of Physics, King's College London, Strand, London
WC2R 2LS, UK

Chappell, G.J.
Department of Mathematics, King's College London, Strand, London WC2R 2LS, UK

Clarkson, T.G.
Department of Electronic and Electrical Engineering, King's College London, Strand,
London WC2R 2LS, UK

Cox, P.
AEA Technology, Culham Laboratory, Oxfordshire OX14 3DB, UK

Haynes, P.
AEA Technology, Culham Laboratory, Oxfordshire OX14 3DB, UK

Hines, E.L.
Department of Engineering, University of Warwick, Coventry CV4 7AL, UK

Hinton, C.
Department of Physics, King's College London, Strand, London  WC2R 2LS, UK

Hutchins, D.A.
Department of Engineering, University of Warwick, Coventry CV4 7AL, UK

Jalel, N.A.
Industrial Control Centre, Faculty of Engineering and Science, The Polytechnic of Central
London, 115 New Cavendish Street, London W1M 8JS, UK

Kodaira, K.
Systems Development Department, Sumitomo Heavy Industries Ltd,
2-4-15 Yato-Cho, Tanashi-City, Tokyo 188, Japan

Lee, J.
Department of Mathematics, King's College London, Strand, London WC2R 2LS, UK

Leigh, J.R.
Industrial Control Centre, Faculty of Engineering and Science, The Polytechnic of Central
London, 115 New Cavendish Street, London W1M 8JS, UK

Mertzanis, E.C.
Advanced Computer Architecture Group, Department of Computer Science,
University of York, Heslington, York YO1 5DD, UK

Mirzai, A.R.
Department of Computer Engineering, University of Science and Technology, Teheran,
Iran

Mottram, J.T.
Department of Engineering, University of Warwick, Coventry CV4 7AL, UK

Nakata, H.
Systems Development Department, Sumitomo Heavy Industries Ltd,
2-4-15 Yato-Cho, Tanashi-City, Tokyo 188, Japan

Nicholson, H.
Emeritus Professor, Department of Control Engineering, University of Sheffield,
Sheffield, UK

Roach, C.
AEA Technology, Culham Laboratory, Oxfordshire OX14 3DB, UK

Shumsheruddin, D.
School of Computer Science, University of Birmingham, Edgbaston,
Birmingham B15 2TT, UK

Smith, M.
AEA Technology, Culham Laboratory, Oxfordshire OX14 3DB, UK

Takamura, M.
Systems Development Department, Sumitomo Heavy Industries Ltd,
2-4-15 Yato-Cho, Tanashi-City, Tokyo 188, Japan

Taylor, J.G.
Director, Centre for Neural Networks, Department of Mathematics, King's College
London, Strand, London WC2R 2LS, UK

Todd, T.
AEA Technology, Culham Laboratory, Oxfordshire OX14 3DB, UK

Toulson, D.L.
Department of Physics, King's College London, Strand, London
WC2R 2LS, UK

Trotman, D.
AEA Technology, Culham Laboratory, Oxfordshire OX14 3DB, UK

Tsang, E.P.K.
Department of Computer Science, University of Essex, Wivenhoe Park,
Colchester CO4 3SQ, UK

Wang, C.J.
Department of Computer Science, University of Essex, Wivenhoe Park,
Colchester CO4 3SQ, UK

Wright, R.E.
Wright Associates, 4 Trenton Close, Frimley, Camberley, Surrey GU16 5LZ, UK

# Learning Algorithms – theory and practice

Norman L. Biggs

London School of Economics
London, UK

## Abstract

Computational Learning Theory is a recently-developed branch of mathematics which provides a framework for the discussion of experiments with learning machines, such as artificial neural networks. The basic ideas of the theory are described, and applied to an experiment involving the comparison of two learning machines. The experiment was devised so that the results could also be compared with those achieved by a human subject, and this comparison raises some interesting questions.

## 1 Introduction

The dramatic increase in the use of artificial neural networks for practical purposes, some of them life-critical, highlights the need for firm foundations in this field. The mathematical study of Computational Learning Theory has developed in parallel with the study of neural networks, and it is now able to contribute to this enterprise. Although the subject is very new, it draws heavily on more established (but also fairly recent) areas of mathematical research, such as Complexity Theory and the Theory of Nonparametric Inference. An introduction to the mathematical theory can be found in the recent book by Martin Anthony and the author [1].

In this paper I shall outline the basic ideas of Computational Learning Theory and, for the purposes of illustration, describe how these ideas can be applied in a simple experimental situation. The 'experiment' involves learning to recognise a geometrical design, and it has some interest in its own right. Although it is very primitive, it throws light on the relative performance of some mechanistic learning algorithms and gives rise to some thought-provoking comparisons with learning by a human subject.

## 2 Computational learning theory

The basic object which we study in Computational Learning Theory is a 'machine' which takes as its input an *example* and outputs a single bit (1 or 0, YES or NO). For our purposes, an example is a string of $n$ bits, representing a pre-processed or coded form of some actual example. Thus it might be a photographic image, coded so that the bits represent black or white pixels, or it might be a set of integer values describing a piece of economic data, coded using the usual binary representation of integers. It is presumed that the set of examples can be split into positive and negative examples, according to some specific concept; for instance, the photographs of beautiful people might be positive examples and the photographs of the rest of us negative examples, or the indications of economic growth might be positive examples and the indications of stagnation negative examples. Formally a *concept* is just a function $t : \{0,1\}^n \to \{0,1\}$, and $y \in \{0,1\}^n$ is a *positive example* of $t$ if $t(y) = 1$, and a *negative example* if $t(y) = 0$.

There are two sets of concepts involved in a typical learning situation. First there is the set of concepts which the teacher might wish to impart, and secondly there is the set of concepts which the learner, or 'machine', is capable of recognising. In reality, the first set could be very large indeed, whereas the second is severely restricted by the architecture of the machine. We shall think of the machine as being able to take on a set $\Omega$ of *states*. If an example $y$ is presented to the machine in a given state $\omega \in \Omega$, then it will output a single bit $h_\omega(y)$, representing its attempt to classify the example. The function $h_\omega : \{0,1\}^n \to \{0,1\}$ is the *hypothesis* of the machine in state $\omega$, and the set of all such functions $\{h_\omega \mid \omega \in \Omega\}$ is its *hypothesis space*. The aim of the learning process is to arrange that the machine takes on a state $\omega$ in which its hypothesis is likely to be a good approximation to a specific concept $t$.

We shall suppose that the information presented to the machine is a sequence of examples, together with the bits representing their classification. Formally, a *training sample* $s$ is just a sequence

$$s = (x_1, b_1), (x_2, b_2), \ldots, (x_m, b_m),$$

where, for $i = 1, 2, \ldots, m$ we have $x_i \in \{0,1\}^n$ and $b_i \in \{0,1\}$. If $b_i = t(x_i)$ $(1 \leq i \leq m)$ then we say that $s$ is a training sample for the *target concept* $t$. We may think of a training sample $s$ as being presented to a machine by a teacher who reveals the desired classification of each example. The aim is to 'train' the machine, using a *learning algorithm* $L$, so that, after examining a training sample $s$ for a target concept $t$, its hypothesis $h_\omega = L(s)$ is likely to be a fairly good approximation to $t$. It must be stressed that the information provided by the teacher is a only a partial definition of $t$; specifically, the values of $t$ on the $m$ examples in the training sample. The final hypothesis $h_\omega$ is a function defined on the entire example space, and only in very exceptional circumstances will it be true that $h_\omega = t$. In order to express these ideas more precisely we require a probabilistic framework for the discussion of concepts and examples, and this will be outlined in Section 5.

When the learning algorithm $L$ has the property that it always produces a hypothesis which agrees with the target concept on the examples which occur in the training sample, then we say that $L$ is *consistent*. The algorithms discussed in this paper all have this property. In mathematical terms, $L$ is consistent if $L(s)$ is an extension to the entire space $\{0,1\}^n$ of the partial function defined by the training sample $s$.

## 3 A logical learning algorithm

As a simple example of a learning algorithm, I shall describe a procedure based on elementary logical principles. A discussion of this method in theoretical terms was first given by Rivest in 1987 [8].

Let $A$ be a subset of $B_n$, the set of all functions from $\{0,1\}^n$ to $\{0,1\}$. A function $t \in B_n$ is said to be a *decision list over A* if there are functions $f_1, f_2, \ldots, f_r$ in $A$ and $c_1, c_2, \ldots, c_r$ in $\{0,1\}$ such that $t(y)$ is given by the following rule.

$$t(y) = \begin{cases} c_j, & \text{if } j = \min\{i \mid f_i(y) = 1\} \text{ exists}; \\ 0, & \text{otherwise.} \end{cases}$$

This simply means that, for a given example $y$, the function values $f_1(y)$, $f_2(y)$, and so on, are evaluated in turn, until we find the first $j$ for which $f_j(y) = 1$; then we set $t(y) = c_j$. If there is no such $j$ we set $t(y) = 0$. We denote the set of all functions in $B_n$ which can be represented by a decision list over $A$ by $DL(A)$.

It is fairly easy to envisage a machine whose hypothesis space is the set of decision lists $DL(A)$, where each 'atomic decision' is one of the functions in $A$. A state $\omega$ of such a machine is a finite sequence

$$\omega = ((f_1, c_1), (f_2, c_2), \ldots, (f_r, c_r))$$

of functions and bits, as described in the previous paragraph. If $A$ is large enough, then the hypothesis space is the entire space $B_n$; in other words the machine is capable of representing *any* target function. Of course, there are substantial problems about the size of the machine required, and these problems are among the major concerns of Computational Learning Theory.

We shall describe a simple learning algorithm which finds a function (hypothesis) in $DL(A)$ consistent with a given training sample, if one exists. We give only the basic idea here: full details, with proofs, may be found in [1] and [8].

The algorithm is as follows. Let $s$ be a training sample consisting of $m$ labelled examples $(x_i, b_i)$ $(1 \le i \le m)$. At each step in the construction of the required decision list some of the examples have been deleted, while others remain. The procedure is to run through $A$ seeking a function $g$ and a bit $c$ such that, for all remaining examples $x_i$, whenever $g(x_i) = 1$ then $b_i$ is the constant boolean value $c$. In other words, $g$ and $c$ provide the correct decision for the remaining examples which satisfy $g$. The pair $(g, c)$ is selected as the next term in the decision list, and all the examples satisfying $g$ are deleted. The procedure is then repeated until either there are no remaining examples, or no pair $(g, c)$ can be found. It is easy to see that, if $s$ is a training sample for a function which is in $A$, then the latter event cannot happen. In other words, the algorithm is successful whenever the possibility of success exists. Also, the algorithm is clearly consistent: it produces a hypothesis which is correct on the examples in the training sample.

We shall use the algorithm in the case when $A = M_n$, the set of monomial functions of $n$ boolean variables. Explicitly, a function in $M_n$ is represented by a formula

$$\lambda_{i_1} \lambda_{i_2} \ldots \lambda_{i_k},$$

where $k \le n$ and each $\lambda$ is a *literal*, that is, one of the objects usually denoted in Boolean algebra by the symbols $x_1, \bar{x}_1, \ldots, x_n, \bar{x}_n$.

It is well-known that every function in $B_n$ can be represented by a disjunction of monomials: this is the DNF (disjunctive normal form) of the function. Furthermore, it is easy to see that a disjunction can be represented as a decision list, since the decision list with terms $(f_1, 1), (f_2, 1)$ defines the same function as the disjunction $f_1 \vee f_2$. It follows that every function can be represented by a decision list of monomials. For the sake of definiteness we shall assume that the monomials are arranged in order of increasing number of literals, and the learning algorithm proceeds by testing them in this order.

At this point it is worth remarking that the decision list 'machine' is a serial processor, in a very extreme sense. Also, its learning algorithm is entirely based on logical principles.

## 4 A neural learning algorithm

By way of contrast to the logic-based algorithm described in the previous section, we shall now consider the Perceptron Learning Algorithm, which was originally conceived as a heuristic description of real neural processes. This classic algorithm was studied in the 1950's and 1960's by Rosenblatt [9], Minsky and Papert [7], and others, and interest in it has recently been revived by the advent of Computational Learning Theory [1, 2, 3, 6].

The relevant hypothesis space is the set of *linear threshold functions* defined as follows. For each $(n + 1)$-tuple $\omega = (\alpha_1, \alpha_2, \ldots, \alpha_n, \theta)$ of real numbers the function $h_\omega$ is defined by

$$h_\omega(y_1, y_2, \ldots, y_n) = \begin{cases} 1, & \text{if } \alpha_1 y_1 + \ldots + \alpha_n y_n \leq \theta, \\ 0, & \text{otherwise.} \end{cases}$$

We refer to $\alpha_1, \alpha_2, \ldots, \alpha_n$ as *weights* and $\theta$ as a *threshold*. The terminology arises because the machine whose hypothesis space is the set of such functions is usually thought of as a set of $n$ input nodes, linked by arcs with weights $\alpha_1, \alpha_2, \ldots, \alpha_n$ to a single active output node with threshold $\theta$.

This is the very simplest artificial neural network, and it is extremely unlikely that a given function in $B_n$ is in its hypothesis space. On the other hand, it is fairly likely that there will be such a function which is consistent with a training sample of length $m$, provided that $m$ is comparable in size with $n$. This is because there are $2^n$ possible examples, and prescribing the values of a function on only about $n$ of them is not very restrictive. This observation justifies the study of learning algorithms, in our sense, even for this very simple machine.

We shall use a discrete version of the Perceptron Learning Algorithm, in which the threshold value $\theta$ is some large positive integer $N$ remaining fixed throughout. Initially all the weights are set equal to zero, and they are always integers. The algorithm is based on the Hebbian principle of changing the weights by a small amount in response to stimuli which are classified incorrectly. More precisely, suppose the current state is $\omega = (\alpha_1, \alpha_2, \ldots, \alpha_n, N)$ and the next labelled example presented in the training sample is $(y, b)$. If $h_\omega(y) = b$ the example is classified correctly, and no change to the state is made. If the example is negative, but is classified by the machine as positive (that is, if $b = 0$ but $h_\omega(y) = 1$), then we reduce the weights on the arcs which are active for that input. That is, we define a new set of weights by the rule

$$\alpha'_i = \begin{cases} \alpha_i - 1, & \text{if } y_i = 1, \\ \alpha_i, & \text{if } y_i = 0. \end{cases}$$

On the other hand, if $b = 1$ but $h_\omega(y) = 0$, then we increase the weights on the active arcs according to the rule

$$\alpha_i' = \begin{cases} \alpha_i + 1, & \text{if } y_i = 1, \\ \alpha_i, & \text{if } y_i = 0. \end{cases}$$

The fundamental result about this algorithm is the so-called perceptron convergence theorem. Roughly speaking, this says that, provided the target concept is a linear threshold function and $N$ is sufficiently large, the algorithm cannot run indefinitely. This means that we can use the algorithm to find a hypothesis consistent with a given training sample, by repeatedly running through the training sample until a run is made in which the weights remain unchanged throughout. The theorem guarantees that this will happen eventually, and when it does the final hypothesis is necessarily consistent with the training sample. As we remarked above, if the training sample is of a size comparable with the number of bits in each example, then there is a good chance that a linear threshold function consistent with the sample exists, and the algorithm will therefore find it.

There are practical problems associated with the perceptron learning algorithm, related to the running time required to produce a consistent hypothesis. It is known [7,9] that this depends on the minimum 'gap' between positive and negative examples in the training sample, regarded as a set of points in $n$-dimensional space. Since it is possible (see [6]) to construct a sequence $\{t_n\}$, where $t_n$ is a concept defined on $\{0,1\}^n$, for which this gap decreases exponentially as a function of $n$, the running time may become impractically large.

## 5 Stochastic concepts

In real-life situations there are two aspects of training which we have overlooked thus far. First, the examples used for training purposes are not, in general, all equally likely. Secondly, the distinction between positive and negative examples is often not clear-cut: some examples may be classified in either way. These two features are present in the experiment described in the following sections, and we need to extend our definitions in order to provide a framework for discussing the results of the experiment.

The basis for this extension is the idea that, for each labelled example $(x, b)$, there is a certain probability of occurrence, denoted by $\tau(x, b)$. More formally, we say that a *stochastic concept* is a probability density function $\tau$ from the product space $\{0,1\}^n \times \{0,1\}$ to the real interval $[0,1]$. If, for every $x$, at least one of $\tau(x,0)$, $\tau(x,1)$ is zero then there is an ordinary concept $t$ defined by

$$t(x) = \begin{cases} 1, & \text{if } \tau(x,0) = 0; \\ 0, & \text{otherwise.} \end{cases}$$

Thus the notion of a stochastic concept extends the ordinary deterministic one. (Note that if both $\tau(x,0)$ and $\tau(x,1)$ are zero, it does not matter how we define $t(x)$, since, with probability one, $x$ does not occur.) A more general treatment of stochastic concepts may be found in reference [4].

When we are given a stochastic concept $\tau$, there is a corresponding probability $\tau^m$ defined on the set of training samples of length $m$: if $s = ((x_1, b_1), \ldots, (x_m, b_m))$, then

$$\tau^m(s) = \tau(x_1, b_1) \ldots \tau(x_m, b_m).$$

This provides the mathematical basis for statements about the 'likelihood' of a some specified outcome of a learning algorithm. For instance, when we say that some particular outcome is highly probable, we mean that its probability with respect to $\tau^m$ is close to 1.

The most obvious new feature of this framework is that a training sample containing the same example classified both positively and negatively may have non-zero probability of occurring. Clearly it is impossible to find a (deterministic) hypothesis which is consistent with such a training sample. Nevertheless, there are circumstances where it is not unreasonable to expect that a consistent learning algorithm will work well, even when the training sample arises from a stochastic concept. This is because the probability of a 'contradiction' in a training sample of length $m$ will be very small, provided that $m$ is comparable with $n$ and there are not too many examples $x$ for which both $\tau(x, 0)$ and $\tau(x, 1)$ are both non-trivial. Thus in practice it may be highly probable that there are no contradictions in the training sample, and we can use a consistent learning algorithm without serious problems.

## 6 An experiment

In order to compare and contrast the practical aspects of the 'logical' and 'neural' approaches to learning, a small experiment was devised. (I am grateful to Dimitrios Tsoubelis for carrying out the programming and practical work.) The idea was to formalise the task of learning to distinguish between positive and negative examples of a geometrical design. In fact, the geometrical features were largely immaterial in the context of learning by machines, since the examples were coded into bit-strings, and the coding did not necessarily reflect the geometric structure. But the geometric features allowed the same tests to be carried out with a human subject, and the results of that experiment will be discussed in the following section.

In order to model a realistic situation, in which the examples occur with differing probabilities, and the distinction between positive and negative examples is blurred to a greater or lesser extent, the training sample was generated by a combination of random processes. This amounts to saying that the target concept is a stochastic concept, as defined in the previous section.

The prototype for the design was a square outline together with one of its diagonals; this was realised by five bands of 16 white pixels with a black background. We shall refer to the 80 pixels which are white in the complete design as the *design pixels*.

The first stage in the generation of a training sample was to decide, using a random equiprobable choice at each step, whether each example was to be positive or negative. Positive examples were then generated by deleting (changing from white to black) design pixels at random; each pixel had probability $p_0$ of being deleted, and the deletions were independent. The value of $p_0$ was 'small', so that a positive example was a slightly defective version of the prototype. Negative examples were generated by a two-stage process. First, one of the five segments was selected at random. each segment having probability 1/5 of being selected. Then design pixels were deleted from the selected segment with probability $p_1$ and from the other four segments with probability $p_2$. In general, $p_1$ was chosen to be 'large', with the intention that a negative example should be a seriously defective version of the prototype. However, the distinction between 'seriously defective' and 'slightly defective' could be blurred by choosing the probabilities $p_0, p_1, p_2$ appropriately.

The difficulty of the distinction depends on the values chosen for $p_0$, $p_1$, and $p_2$. If $p_1$ is large compared with $p_0$, then a negative example can be expected to contain one segment which is much more 'faded' than any segment in a typical positive example; on the other hand if $p_0 = p_1 = p_2$, then there is no distinction between positive and negative examples. Furthermore, in order to ensure that the distinction between positive and negative examples cannot be based simply on counting pixels, the probabilities must be chosen so that the expected number of pixels deleted is roughly the same in both cases. For the positive examples it is $80p_0$, and for the negative examples it is $16p_1 + 64p_2$. So, if we define

$$\delta = 80p_0 - 16p_1 - 64p_2,$$

we need to ensure that $|\delta|$ is small. The following sets of values of $(p_0, p_1, p_2)$ were used in the experiment; they satisfy the condition $|\delta| \leq 5$ and in each case the distinction between positive and negative examples is neither too blurred nor too clear-cut.

$(0.10, 0.30, 0.05)$, $(0.10, 0.30, 0.10)$, $(0.10, 0.40, 0.05)$, $(0.10, 0.40, 0.10)$,

$(0.15, 0.45, 0.05)$, $(0.15, 0.45, 0.10)$, $(0.15, 0.60, 0.05)$, $(0.15, 0.60, 0.10)$,

$(0.20, 0.60, 0.05)$, $(0.20, 0.60, 0.10)$, $(0.20, 0.80, 0.05)$, $(0.20, 0.80, 0.10)$.

The first part of the experiment was intended to test the effectiveness of the algorithms in finding a consistent hypothesis. Given $(p_0, p_1, p_2)$. a training sample of 100 labelled examples was generated as described above. Each example was coded as a string of 80 bits in the obvious way. The training sample was then used as the input to a (simulated) monomial decision list machine and a (simulated) perceptron machine, and, using the appropriate learning algorithm in each case, a hypothesis consistent with the training sample was sought.

Since the training sample is generated by a stochastic concept it may, in theory, contain contradictions, and when this happens it is impossible to find a consistent hypothesis. (It must be remembered that these machines produce deterministic hypotheses.) However, contradictions are unlikely to happen in practice when the parameters are those used here. In fact, each training sample was checked for contradictions, and none were found in any of the tests described below.

In the case of the decision list algorithm, we aimed to produce a consistent hypothesis in which $k$, the maximum degree of the monomials, is as small as possible. Somewhat surprisingly, it turned out that in all cases it was possible to find a consistent decision list with $k = 1$. In other words, the atomic decisions were based simply on whether a single pixel was white or black. The length of the decision list varied from about 30 to about 40.

In the case of the perceptron, even in the deterministic case it is not guaranteed that there will be a linear threshold function consistent with the training sample, although this is highly probable when we have only 100 examples, and the number of bits in each example is of the same order. Here too, a consistent hypothesis was always found in the tests. As a check on the method, some tests were repeated with the same training sample, but with the value of the threshold $N$ doubled. If all is well, we should expect the resulting final values of the weights to be doubled also, and in all cases this turned out to be so.

As remarked above, there are heuristic arguments which help to explain why the search for a linear threshold function consistent with the training sample was always successful in this experiment. (In fact this was found to be so even when there was no meaningful distinction between positive and negative examples ($p_0 = p_1 = p_2$), which indicates the dangers of claiming extraordinary powers for neural networks.) We also remarked in Section 4 that the number of iterations needed depends on the 'gap' between the positive and negative examples. When the training sample is generated by a random process, the gap may occasionally be rather small, and the running time will then increase substantially. This phenomenon was observed in several tests.

The second part of the experiment was intended to test the hypothesis produced in the first part to see how well it represented the target stochastic concept. The values $(p_0, p_1, p_2)$ determine the probability $\tau(x, b)$ that each labelled example occurs at each step in the generation of the initial training sample. For example, the probability that an example is positive is $1/2$, and the probability of deleting a given number of pixels is binomially distributed with parameters $n = 80$ and $p_0$; consequently we have

$$\tau(x, 1) = \frac{1}{2} p_0^{\delta(x)} (1 - p_0)^{80 - \delta(x)},$$

where $\delta(x)$ is the total number of pixels which must be deleted from the complete design to obtain $x$. The probability of a negative example is more complicated, but it can be written down quite easily. Thus, if required, we have an explicit description of the target stochastic concept $\tau$ determined by $p_0$, $p_1$, and $p_2$.

When the initial training sample is presented as input to one of the algorithms, the result is the construction of a (deterministic) hypothesis $h$. As remarked above, we can be fairly confident that a hypothesis $h$ which agrees with the training sample will be found, because it is unlikely that the training sample contains contradictions. Another training sample (known for clarity as the *testing sample*), consisting of 100 further labelled examples $(x, b)$, is then generated using the same values of $p_0$, $p_1$, and $p_2$ – that is, the same $\tau$. When the values $h(x)$ are compared with the corresponding bits $b$, the results can be tabulated as follows.

| $b$ | $h(x)$ | % |
|---|---|---|
| 0 | 0 | $n_{00}$ |
| 0 | 1 | $n_{01}$ |
| 1 | 0 | $n_{10}$ |
| 1 | 1 | $n_{11}$ |

The *success rate* of a learning algorithm, for the given stochastic concept $\tau$, is measured by the percentage of correct classifications which it achieves on the testing sample; that is, $n_{00} + n_{11}$ in the table above.

The following tables show the success rates obtained for the decision list (DL) and perceptron (P) algorithms for the chosen values of $(p_0, p_1, p_2)$.

| $p_0$ | $p_1$ | $p_2$ | DL | P |
|------|------|------|------|------|
| 0.10 | 0.30 | 0.05 | 76 | 59 |
| 0.10 | 0.30 | 0.10 | 71 | 70 |
| 0.10 | 0.40 | 0.05 | 77 | 59 |
| 0.10 | 0.40 | 0.10 | 80 | 66 |
| 0.15 | 0.45 | 0.05 | 89 | 55 |
| 0.15 | 0.45 | 0.10 | 78 | 55 |
| 0.15 | 0.60 | 0.05 | 76 | 65 |
| 0.15 | 0.60 | 0.10 | 75 | 70 |
| 0.20 | 0.60 | 0.05 | 81 | 57 |
| 0.20 | 0.60 | 0.10 | 85 | 61 |
| 0.20 | 0.80 | 0.05 | 85 | 56 |
| 0.20 | 0.80 | 0.10 | 88 | 70 |

Perhaps the most striking thing about these results is that DL outperforms P in all cases. It might be argued that the perceptron, having just one active node, is the simplest form of neural net or parallel processor, and it should not be compared with more sophisticated algorithms. But we have already remarked that the decision lists which arise in the experiment involve atomic decisions based only on single pixels, and so in a sense DL, as used here, is the simplest form of serial processor.

## 7 The experiment with a human subject

As we have already noted, the description of the learning task in terms of a geometrical design is largely immaterial when the examples are coded as bit-strings and presented to a machine. One reason for using such a task was that it allowed the performance of the machines to be compared with that of a human subject. We produced a visual display of the training sample of 100 defective versions of the geometrical design (together with the target classification of each one) and presented it to a human subject. The subject was allowed to study the display and asked to form a hypothesis consistent with this training sample. The testing sample was then presented (without the target classification), and the subject was asked to classify these examples on the basis of the hypothesis formed in the first stage.

When a human subject was given a single test run of the experiment the results were roughly comparable with those for the mechanistic algorithms, although there was considerable variability between subjects. Subjects with a scientific background reported afterwards that they had attempted to formulate quite complex hypotheses, and this could lead to disastrous results. Non-scientific subjects were happy to work with a vague and intuitive hypothesis, and would usually have a success rate of over 50%. For example, with $p_0 = 0.20, p_1 = 0.60$ and $p_2 = 0.05$ one subject achieved a success rate of 57% on the first test run; coincidentally the same success rate was achieved by the perceptron algorithm, although the decision list algorithm did significantly better, with 81%.

However, the behaviour of human subjects over a number of tests differed quite noticeably from that of the machines. The most obvious reason for this is that human beings have the ability to acquire concepts in a cumulative fashion. In the context of our experiment this affected their behaviour in two ways.

First, a human subject who is shown the visual display will recognise immediately, that each example consists of five separate entities, the five segments of the design. Such recognition is the result of prior organisation which enables human's visual system to 'see' lines. There is much debate about whether this ability is learned or genetically determined ('hard-wired'), but it seems likely that learning is involved to some extent. Occasionally people who have been blind from birth are given sight in later life, and these people cannot immediately make sense of their visual input [5]. Of course, it is not at all obvious how such prior learning at a perceptual level could be simulated mechanically by the means at our disposal. But we could simulate the end-product of the process, simply by splitting the input into five parts, each of 16 consecutive bits, corresponding to the five segments of the design. This device would enable us to concentrate on other, possibly more amenable, aspects of the human learning process.

The second way in which the behaviour of human subjects differs from that of the machines is that the humans were able to 'learn' at a higher level when the test was repeated several times. In one case the performance of a human subject improved after several tests, to the point where a success rate of nearly 100% was achieved. Thus it appears that human subjects progressed from being 'naive' to being 'sophisticated'. Naive subjects were new to the experiment and based their hypothesis solely on the training sample for the current test. Sophisticated subjects had done several tests, and had begun to 'understand' the nature of the distinction between positive and negative examples. The improved success rate of sophisticated subjects is not simply due to their having seen more examples; it seems to be more convincingly explained by a kind of 'two-tier' model of learning, as suggested by Valiant [10], involving a combination of supervised and unsupervised learning. It would be instructive to devise a mechanistic procedure which would reproduce the unsupervised learning displayed by humans in our experiment, and which could be combined with supervised learning algorithms like DL and P. As remarked above, we should have to allow the unsupervised learning component to use the partitioned form of the input, because this corresponds to the organisation or 'pre-processing' carried out by the human visual system.

## References

[1] M. Anthony and N.L. Biggs, *Computational Learning Theory: an Introduction*, Cambridge University Press, 1992.

[2] P.L. Bartlett and R.C. Williamson, Investigating the distribution assumptions in the pac model, *Proceedings of the 1991 Workshop on Computational Learning Theory*, Morgan Kauffman, San Mateo, 1991.

[3] E. B. Baum, The perceptron algorithm is fast for non-malicious distributions, *Neural Computation*, 2 (1990) 249-261.

[4] A. Blumer, A. Ehrenfeucht, D. Haussler and M. Warmuth, Learnability and the Vapnik-Chervonenkis dimension, *Journal of the ACM*, 36 (1989) 929-965.

[5] R. L. Gregory and J. G. Wallace, Recovery from early blindness: a case study, *Exp. Psychol. Soc. Monographs No. 2*, Cambridge 1963.

[6] N. Littlestone, Learning quickly when irrelevant attributes abound, *Machine Learning*, 2 (1988) 245-318.

[7] M. Minsky and S. Papert, *Perceptrons*, MIT Press, expanded edition 1988.

[8] R. Rivest, Learning decision lists, *Machine Learning*, 2 (1987) 229-246.

[9] F. Rosenblatt, *Principles of Neurodynamics*, Spartan Books, 1962.

[10] L. G. Valiant, A view of computational learning theory, *NEC Research Symposium: Computation and Cognition* (ed. C. W. Gear), SIAM, 1991.

# A Generic Neural Network Approach For Constraint Satisfaction Problems

Dr. E. P. K. Tsang
Dr. C. J. Wang

Department of Computer Science, University of Essex
Wivenhoe Park, Colchester, UK

## Abstract

The Constraint Satisfaction Problem (CSP) is a mathematical abstraction of the problems in many AI application domains. In many of such applications timely response by a CSP solver is so crucial that to achieve it, the user may be willing to sacrifice completeness to a certain extent. This paper describes a neural network approach for solving CSPs which aims at providing prompt responses. The effectiveness of this model, which is called GENET, in solving CSPs with binary constraints is demonstrated by a simulator. Although the completeness is not guaranteed, as in the case of most of the existing stochastic search techniques, solutions have been found by the GENET simulator in all of our randomly generated problems tested so far. Since the neural network model lends itself to the VLSI implementation of parallel processing architectures, the limited number of cycles required by GENET to find the solutions for the tested problems gives hope for solving large CSPs in a fraction of the time required by conventional methods.

## 1    Introduction

Constraint satisfaction is the nature of the problems in many AI application domains. For instance, line labelling in vision [1], subproblems in temporal reasoning [2, 3], and resource allocation in planning and scheduling [4, 5] can all be formulated as Constraint Satisfaction Problems (CSPs). A CSP is defined as a triple $(Z, D, C)$, where $Z$ is a finite set of variables, $D$ is a set of domains, one for each variable, and $C$ is a set of constraints. Each constraint in $C$ restricts the values that one can assign to a set of variables simultaneously. The task is to assign one value per variable, satisfying all the constraints in $C$ [6].

To make discussion easier we shall first define a few terms here. The assignment of a value to a variable is called a *label*. A (possibly empty) set of labels is called a *compound label*. A compound label with $k$ labels in it is called a *k-compound label*. A constraint is *n-ary* if it applies to $n$ variables. A *binary CSP* is a CSP with unary and binary constraints only.

This research is motivated by the need to produce prompt solutions to CSPs.

There are at least two reasons for such need:

## (A) Time is limited

In some applications the user has very limited time to solve the problems. For example, in allocating resources such as manpower and equipments to jobs in a workshop, one could be dealing with a very dynamic situation. For instance, jobs may be added, cancelled or given priority, and resources could become unavailable, etc. If these changes happen frequently one may not be allowed too much time for scheduling and re-scheduling.

Furthermore in many scheduling problems the user has to consider a large number of combinations of constraints. Sometimes not all the constraints can be satisfied and, therefore, the user has to consider relaxing some of them. To enable him/her to evaluate the different situations the real time response of a CSP solver is crucial.

## (B) Time determines cost

There are applications where cost is a function of the response time. Imagine sitting in the control room of a container port and allocating storage locations to containers unloaded from container vessels. Each minute's delay in clearing the containers from a busy port could cost hundreds of pounds. Therefore one would like to decide as soon as possible where to store the containers.

Besides, in a catastrophic incident, such as a natural disaster or an accident in a nuclear power station, one of the coordinator's tasks may be to allocate resources, such as rescuing teams and scarce equipments, to different sites. The coordinator's reaction time may directly affect the amount of loss in terms of life and death and the amount of properties and equipments being destroyed.

The majority of existing work in CSP has been focused on problem reduction and heuristic search [6, 7]. A few techniques in such categories have been used in the constraint programming language CHIP. For symbolic variables, CHIP uses the Forward Checking algorithm together with the Fail First Principle (FC-FFP) heuristic [7]. CHIP's strategy has worked reasonably well in certain real life problems [8, 9, 10]. Unfortunately, CSPs are NP-hard in nature. The size of the search space in a CSP is typically $O(d^N)$, where $d$ is the size of each domain in $D$ (assuming, for simplicity, that all domains have the same size) and $N$ is the number of variables in the problem. For CSPs with, say, $N = 10000$ variables and $d = 50$, the search space will be so huge ($50^{10000}$) that even a good heuristic algorithm may not be able to produce any answer within a tolerable period of time.

Speed could be gained in CSP solvers by using parallel processing architectures. Indeed, Saletore and Kale show that to a certain extent, linear speedup is achievable by using parallel architectures [11]. However, as Kasif points out, problem reduction (which CHIP uses) is inherently sequential, and it is unlikely that a CSP can be solved in logarithmic time by using only a polynomial number of processors [12].

Neural Network (NN) techniques give hope to a higher degree of parallelism in solving CSPs. Although NNs normally do not guarantee to find solutions even when some exist (as they could be trapped in local minima or local maxima), their stochastic hill-climbing nature may also lead them to producing solutions much more quickly (if local minima are avoided). Hopfield and Tank's work on the Travelling Salesman Problem is one such example [13]. The attempt to apply neural network techniques to solve CSPs has led to the discovery of the Heuristic Repair Method, which is based on a heuristic called Min-conflict [14, 15]. The Heuristic Repair Method can solve the million-queens problem in minutes.

Other work in applying *connectionist* approaches to CSPs can be found in [16, 17, 18 and 19]. Swain and Cooper [16] propose to use a connectionist approach to problem reduction (through maintaining arc-consistency [16]), and Cooper [17] applies this technique to graph matching. Guesgen [18] extends Swain and Cooper's approach to facilitate the generation of solutions from the converged network. However, generating solutions from the converged network is far from trivial. Collin et al [19] point out that even for relatively simple constraint graphs, it is difficult to find a general parallel computational model which guarantees completeness. Besides, all these approaches require massively parallel architectures.

The major problem with neural network approaches in general is the danger of the network settling in local minima, which prevents the problem solver from finding solutions when some exist. Our analysis and experiments have shown that the Heuristic Repair Method is only effective for CSPs for which there exist a large number of solutions, such as the N-queens problem with large N [20].

In this paper, we describe a generic neural network approach for solving CSPs with binary constraints. The model that we propose is called GENET. The effectiveness of GENET is demonstrated by a simulator, which can generate connected networks dynamically for binary constraint problems, and simulate the network convergence procedure. In case the network falls into local minima, a heuristic learning rule will be applied to escape from them. The network model lends itself to high degrees of parallelism. The experimental results of applying the GENET simulator to randomly generated CSPs have been very favorable, which indicates that massive parallelism is not necessary to achieve a few orders of magnitude speedup over sequential heuristic search method.

## 2   NETWORK MODEL

A binary constraint problem is represented in GENET in the following way. One node in the NN is used to represent one label. Each node has a binary output - either it is *on* (*active*) or it is *off* (*inactive*). If a node is on, it means the corresponding assignment is being made. The set of all nodes which represents the labels for the same variable forms a cluster. The connections among the nodes are constructed according to the constraints $C$, with an inhibitory connection between incompatible nodes and no connections between compatible nodes. The weight of all the connections are initialized to *-1*. Figure 1 shows a CSP and its network in GENET.

By convention, we shall denote the state of node $i$ as $s_i$, which is either *1* for *on* or *0* for *off*. The weight of the connection between nodes $i$ and $j$ is denoted as $w_{ij}$, which is always a negative integer and initially given the value *-1*. The input to a node is the weighted sum of all its connected nodes' states. Only one node in each cluster will be allowed to turn on. Therefore every state of the network represents an assignment of one value per variable. Any network state in which no two ON nodes are connected represents a solution to the CSP.

## 3    Network Convergence and Escaping Local Minima

The network convergence procedure is as follows. Initially one node in each cluster is randomly selected to turn on. Then, in each convergence cycle, every node calculates its input. In each cluster, the node that receives the maximum input will be turned on, and the others will be turned off. Since there exist only negative connections (representing the constraints in the problem), the winner in each cluster represents a value which, when being assigned to the corresponding variable, violates the fewest

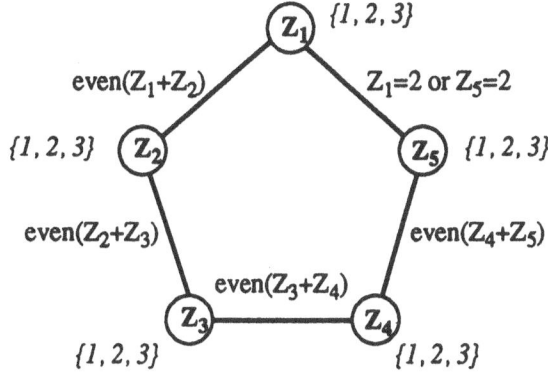

(a) A CSP with 5 variables, $Z_1$, $Z_2$, $Z_3$, $Z_4$ and $Z_5$,
all with the same domain {1, 2, 3}

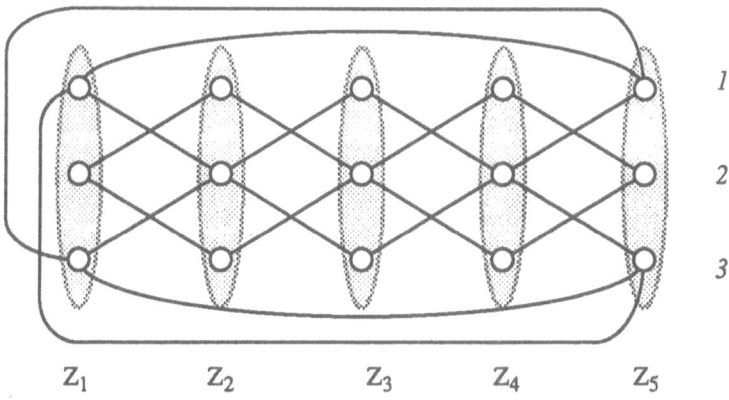

(b) The network corresponding to (a) generated by GENET

Figure 1. A CSP and its corresponding network generated by GENET

constraints. In tie situations, if one of the nodes in the tie was on in the previous cycle, it will stay on. If all the nodes in the tie were off in the previous cycle, a random choice is made to break the tie. This is to avoid chaotic or cyclic wandering of the network states. We have experimented in breaking ties randomly, as is done in the Heuristic Repair Method [15]. But we find this strategy ineffective in problems which have few solutions.

If and when the network settles in a stable state, GENET checks if that state represents a solution. As stated before, a state in which all the active nodes have zero input represents a solution. Otherwise, it represents a local minimum.

When the network settles in a local minimum, the state update rule fails to make alternative choices for variable assignments based on the local information received at each cluster of nodes. It would appear that introducing randomness or noise in the state update rule, in a manner of the simulated annealing as suggested in the literature [21], might help in escaping local minima. However, this will degrade the overall performance drastically, which could make this approach not effective enough for solving any real life problems. Instead, we have adopted a heuristic learning rule which updates the connection weights as follows:

$$w_{ij}(t+1) = w_{ij}(t) - s_i(t) \times s_j(t)$$

where $t$ is a discrete time instance measured in terms of cycles. This learning rule effectively does two things: (a) it continuously reduces the value of the current state until it ceases to be a local minimum, and (b) it reduces the possibility of any violated constraint being violated again. Hopefully, after sufficient learning cycles, the connection weights in the network will lead the network states to a solution. The network convergence algorithm is shown in pseudo codes in Figure 2.

The states of all the nodes are revised in parallel asynchronously in the model. The outer REPEAT loop terminates when a solution is found, or some resource is exhausted, which could mean that the maximum number of cycles have been reached, or the time limit has been exceeded. The resources available vary from application to application.

# 4    Experiments and Evaluation of GENET

In order to test the effectiveness of GENET for solving CSPs, a GENET simulator is constructed. Thousands of tests have been performed on randomly generated and specially designed CSPs to evaluate the performance of GENET. Random problems are generated with the following 5 parameters varied:

$N$    the number of variables;

$D$    the maximum size of domains;

$d$    the average size of individual domains ($d \leq D$) - the introduction of $d$ in addition to $D$ is to allow us to test problems which may have different domain sizes;

$p_1$    the percentage of pairwise variables being constrained - in the generated problem, there are $p_1 \times \dfrac{N \times (N-1)}{2}$ constraints; and

$p_2$    the percentage of compound labels satisfying each binary constraint; e.g. if there exists a binary constraint between variables $X$ and $Y$, and the domain sizes of $X$ and $Y$ are $d_1$ and $d_2$, then $p_2$ of the $d_1 \times d_2$ combinations of labels for $X$ and $Y$ are compatible with each other.

The probability for any two randomly assigned variables being compatible, $p_c$, can be obtained as follows:

$$p_c = 1 - p_1 + p_1 \times p_2$$

The probability of an $N$-compound label violating no constraint is defined as the *tightness* of a problem. With $N$ variables:

$$tightness = \prod_{i=1}^{N-1} p_c^i = p_c^{\frac{N(N-1)}{2}}$$

```
PROCEDURE GENET
BEGIN
   One arbitrary node per cluster is switched ON;
   REPEAT /* network convergence: */
     REPEAT
       Modified := False;
       FOR each cluster DO IN PARALLEL
         BEGIN
           On_node := node which is at present ON;
           Label_Set := the nodes with the maximum input;
           IF NOT (On_node in Label_Set) THEN
             BEGIN
               Modified := True;
               On_node := OFF;
               Switch an arbitrary node in Label_Set to ON;
             END
         END
     UNTIL (NOT Modified); /* the network has converged */
     /* learning: */
     IF (sum of input to all ON nodes < 0) /* in local minma */
         THEN FOR connection c between nodes x and y DO IN PARALLEL
         IF (both x and y are ON)
             THEN decrease the weight of connection c by 1;
   UNTIL (input to all ON nodes are 0) OR (any resource exhausted)
END
```

Figure 2. GENET convergence procedure

A problem is *tight* if its tightness is small. The GENET simulator is tested on problems with varying number of variables, domain sizes and degrees of tightness (by varying $p_1$ and $p_2$). Tests have been focused on tight problems to see how likely GENET is to miss solutions. GENET is given a limit in the number of convergence cycles. If this limit is exceeded before a solution is found, GENET is instructed to report a failure. When GENET reports a failure, the problem is passed to a program which performs a complete search to check if GENET has missed any solution. For all the problems tested, GENET is capable of converging on solutions in solvable problems, and reporting failures in insoluble problems.

The time required by GENET to find solutions is measured by the number of convergence cycles required by the simulator. For CSPs in which $N = 170$, $D = d = 6$, $p_1 = 10\%$ and $p_2 = 85\%$, GENET takes just over 100 cycles to terminate. For a fully parallelized (with $d \times N$ parallelism) VLSI implementation one cycle may take a few hundred nano seconds. Thus, a problem of the aforementioned size can be solved in terms of tens to hundreds of micro seconds. Compared with a sequential heuristic search which would take over 20 hours to solve the same problem, the speedup is in the order of $10^8$. The speedup of such a high order implies that a very high overhead could be affordable if the full parallelism is not supported. In fact, a parallelism of $d$ would give a respectable speedup, as shown below.

Take the above problem for example. Suppose we have only $d$ parallel processors. This would allow us to process one cluster at each processing cycle. To allow for more adverse situations, suppose each processor has only one input port. This would mean that one network convergence cycle had to be completed in $N(N-1)$ processing cycles ($N$ clusters to be processed in order and each cluster needs $N-1$ cycles for adding up the active input from all other clusters). Since $N = 170$, the speedup will be degraded down to the order of $10^4$. Considering this 4 orders of magnitude speedup being achievable at a moderate parallelism ($d$) and with a relatively simple and low cost processor (one input port), we may conclude that the GENET approach indeed fulfills our initial research objectives.

Another interesting finding in our experiments is that the weights in the network never fall below -50 and the total input to a node never fall below -100. The range of these values could be further reduced if floating points instead of integers were used for the weights. This finding is important for the analogue VLSI implementation of the GENET model, as shown below.

## 5  VLSI Implementation

A full discussion of a VLSI implementation of the GENET model is beyond the scope of this paper. In this section, we will simply show the feasibility of such implementation using analogue techniques.

The computation in each GENET convergence cycle can be divided into two parts: (a) the summation of input signals to a node, and (b) the competition within a cluster. The summation can be realized by a linear operational amplifier and the competition can be realized by an analogue channel comparator [22].

Figure 3 shows a schematic diagram for the design of one cluster of nodes, where the vertical box represents the channel comparator with one channel receiving the total input signal of each node in the cluster. One single-bit register is associated to a channel, as shown to the right hand side of the vertical box (labeled D), to store the state of the corresponding node. The channel that receives the highest signal level will be switched on, and consequently its register will be set. All other channels will be switched off and so are their registers.

The state of the registers are also fed back to the channel comparator so that if the channel with the highest signal level already has its associated register set, a signal of *no change* is generated for this cluster. Thus, local minima can be detected. The comparator also checks the highest signal level against a *null* reference signal. If they match, a signal of *no violation* is generated. Henceforth, the solution states can be detected. The tie situation is solved by the natural deviation in the circuitry parameters and the feedback from the state registers.

Clearly the sensitivity of the channel comparator depends on the minimum interval amongst the levels of the signals being compared, and the range of signal

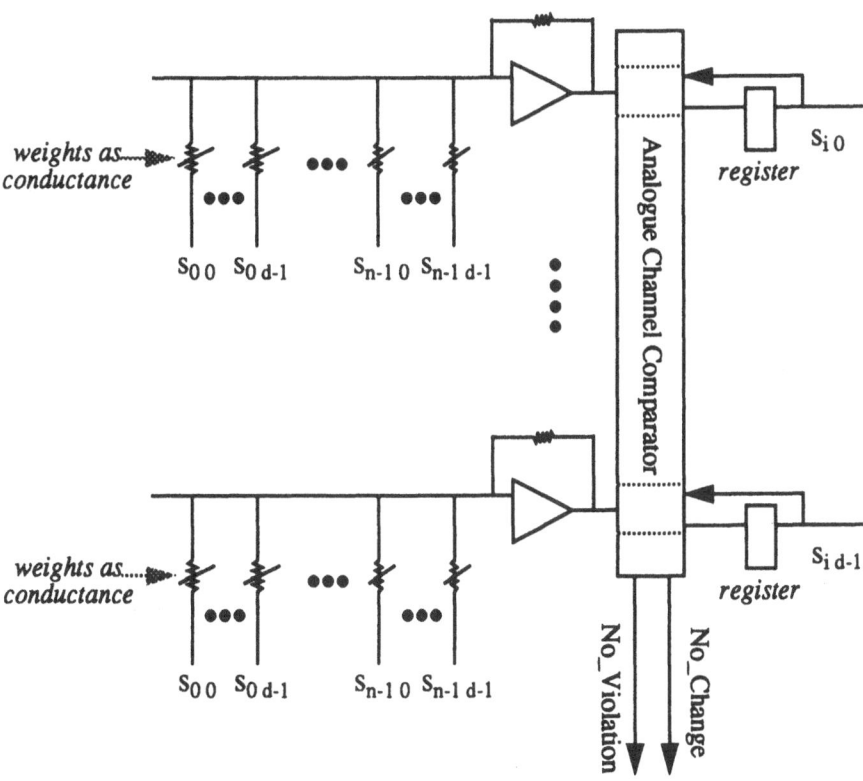

Figure 3. The design of one cluster of nodes, where $s_{ij}$ represents the state of the *j*th node in cluster *i*

levels is limited by the VLSI technology. However this is not a problem with the GENET model because, as mentioned before, the weights and total input to a node are well limited. Firstly, a fine tuning of the weights' granularity can help reduce the total range of the signal levels. Secondly, the saturation at the lower end of the signal levels does not affect the competition. Lastly, should it be necessary, a uniform signal level shift can be applied to solve the problem.

It should be noted that Figure 3 only shows the design principle. The technical detail of a cascadable VLSI architecture for GENET is described in [22], which proposes an even simpler design that does not require modifiable resistors for learning purposes. The weights are stored off chip, the size of the channel comparator is expandable by cascading such neuro-chips vertically, and the number of clusters processed concurrently is expandable by cascading such neuro-chips horizontally.

# 6 Future Work

Research in GENET is on-going. Our current research focus is in the following directions. Firstly, we are evaluating GENET's performance in solving partial constraint satisfaction problems (PCSPs) [23]. One instance of the PCSPs which is useful for scheduling is to minimize the number of constraints violated when the problem is over-constrained. The PCSP is difficult because one can not use constraints to prune off as much of the search space in it as in the standard CSP (because the optimal solution may violate some constraints).

The second major direction that we are investigating is to handle general constraints. For non-binary problems and problems in which the values are not simple ground terms, a multilayer structured network will be required. The number of layers required depends on the complexity of the value structure. We have performed preliminary tests in applying GENET to networks constructed (by the experimenters) for instances of the car-sequencing problem [8, 24]. Like all the other tests so far, GENET finds solutions in those networks. Our approach to the car-sequencing problem is outlined in [20].

Our long term research direction is to fabricate VLSI neuro-chips for constructing CSP solvers for real life applications.

# 7 Concluding Summary

We have pointed out that there are applications in which timely response by CSP solvers is so crucial that a limited degree of sacrifice in completeness by the CSP solver can be justified. For such applications, we have proposed a neuro-network based model called GENET whose aim is to produce solutions fast enough for real time applications. A simulator has been built to demonstrate the effectiveness of GENET. Although completeness in GENET is not guaranteed, the GENET simulator has not yet failed in finding solutions for thousands of randomly generated CSPs which vary significantly in their size and tightness. We have shown the feasibility of building hardware for GENET. The fact that relatively few cycles are needed by the GENET simulator to find solutions gives hope for solving CSPs in a fraction of the time required by conventional techniques.

# Acknowledgment

The authors have benefited from discussions with Dr J. Ford, Dr J. Reynold and Kar-Lik Wong in this research. Jenny Emby has greatly improved the readability of this paper.

# References

[1]   Waltz, D.L., "Understanding line drawings of scenes with shadows", in WINSTON, P.H. (ed.) The Psychology of Computer Vision, McGraw-Hill, New York, 1975, 19-91

[2]   Tsang, E.P.K., "The consistent labelling problem in temporal reasoning", Proc. AAAI Conference, Seattle, July, 1987, 251-255

[3]   Dechter, R., Meiri, I. & Pearl, J., "Temporal constraint networks", Artificial Intelligence, 49, 1991, 61-95

[4]   Stefik, M., "Planning with Constraints (MOLGEN: part 1)", Artificial Intelligence 16, 1981, 111-140

[5]   Prosser, P., "Distributed asynchronous scheduling", PhD Thesis, Department of Computer Science, University of Strathclyde, November, 1990

[6]   Mackworth, A.K., "Consistency in networks or relations", Artificial Intelligence 8(1), 1977, 99-118

[7]   Haralick, R.M. & Elliott, G.L., "Increasing tree search efficiency for constraint satisfaction problems", Artificial Intelligence 14(1980), 263-313

[8]   Dincbas, M., Simonis, H. & Van Hentenryck, P., "Solving car sequencing problem in constraint logic programming", Proceedings, European Conference on AI, 1988, 290-295

[9]   Dincbas, M., Van Hentenryck, P., Simonis, H., Aggoun, A. & Graf, T., "Applications of CHIP to industrial and engineering problems", First International Conference on Industrial and Engineering Applications of AI and Expert Systems, June, 1988

[10]  Perrett, M., "Using Constraint Logic Programming Techniques in Container Port Planning", ICL Technical Journal, May 1991

[11]  Saletore, V.A. & Kale, L.V., "Consistent linear speedups to a first solution in parallel state-space search", Proc. AAAI, 1990, 227-233

[12]  Kasif, S., "On the parallel complexity of discrete relaxation in constraint satisfaction networks", Artificial Intelligence (45) 1990, 275-286

[13]  Hopfield, J. J., and Tank, D.W., "'Neural' Computation of Decisions in Optimization Problems", Biol. Cybern. (52) 1985, 141-152

[14]  Adorf, H.M. & Johnston, M.D., "A discrete stochastic neural network algorithm for constraint satisfaction problems", Proceedings, International Joint Conference on Neural Networks, 1990

[15]  Minton, S., Johnston, M.D., Philips, A. B. & Laird, P., "Solving large-scale constraint-satisfaction and scheduling problems using a heuristic repair method", Proc. AAAI, 1990, 17-24

[16]    Swain, M.J. & Cooper, P.R., "Parallel hardware for constraint satisfaction", Proc. AAAI, 1988, 682-686

[17]    Cooper, P., "Structure recognition by connectionist relaxation: formal analysis", Proc. Canadian AI Conference (CSCSI), 1988, 148-155

[18]    Guesgen, H.W., "Connectionist networks for constraint satisfaction", AAAI Spring Symposium on Constraint-based Reasoning, March, 1991, 182-190

[19]    Collin Z., Dechter, R. & Katz, S., "On the Feasibility of distributed constraint satisfaction", Proc. International Joint Conference on AI, 1991, 318-324

[20]    Wang, C. J., & Tsang, E. T. K., "Solving constraint satisfaction problems using neural networks", Proc. IEE Second International Conference on Artificial Neural Networks, 1991

[21]    Davis, L. (ed.), "Genetic algorithms and simulated annealing", Research notes in AI, Pitman/Morgan Kaufmann, 1987

[22]    Wang, C.J., "A cascadable parallel architecture for GENET", Research Notes, Department of Computer Science, University of Essex, forthcoming.

[23]    Freuder, E.C., "Partial constraint satisfaction", Proc., 11th International Joint Conference on AI, 1989, 278-283

[24]    Parrello, B.D., Kabat, W.C. & Wos, L., "Job-shop scheduling using automated reasoning: a case study of the car sequencing problem", Journal of Automatic Reasoning, 2(1), 1986, 1-42

# A Review of Medical Diagnostic Applications of Neural Networks

G.J. Chappell, J. Lee and J.G. Taylor
King's College, London

## Abstract

Recent years have seen increasingly many attempts to use Neural
Network methods to solve computationally difficult problems in
medical diagnosis.   This paper presents a preliminary review of
applications, noting some problems encountered, along with areas in
which success has been achieved.  It is observed especially that neural
network theory is not always applied with sufficient care.   Some
thoughts are advanced on the future applicability of neural networks in
medical science, noting that priorities differ from those of industry.

Papers dealing with applications of neural networks in medicine seem to be less
concerned than their industrial counterparts with benchmarking against "standard"
methods.   In heavy industrial applications, the advantage sought in automation is
probably nothing other than reduced running costs.   Standard methods have often
been implemented successfully and as such, neural network methods have to show a
substantial performance margin.

Studies of the effectiveness of neural network methods, such as the ANNIE
report [1], have examined neural networks in the industrial context, showing that they
perform well, but no better than the usual statistical classifiers (and some fairly
unusual ones, the comparisons having been near to exhaustive).   In technical terms,
the problems dealt with in the tests are similar to those encountered in medical
physics, involving such things as ultrasound imaging, but it is clear that there are
necessary differences in approach.

In medical physics, less time is available for data processing, partly because the
amount of data is often huge, but also because patient treatment (and perhaps even
life) may depend on having results quickly.  High diagnostic accuracy is essential, but

in many problems, even the opinion of trained experts is variable and attempts at automation using standard methods have so far proved unreliable, especially in imaging.

The difference in attitude may be seen clearly in [2], in which an attempt is made to use electroretinogram signals obtained from exposing patients to a reversing chequerboard pattern as an indication of the scores these patients would have received in a related contrast sensitivity test. Expert prediction of the relationship is shown to be highly variable, despite the belief in theory that the two tests are in fact related, and as such, any method with better performance will be welcomed. Although much thought is given to the design and interpretation of data, the choice of network architecture receives no consideration whatsoever.

Anxiety about the time taken by standard methods of processing medical information is expressed well in a surprisingly entertaining paper [3]. To quote,

> "Currently, this assessment is made by experts skilled in the interpretation of ... potentials according to a visually-based grading scheme. The testing ... requires approximately two hours and because of the labour intensiveness of the analysis methods, the test is currently administered once every 24 hours. The fact that 50 percent of severely head-injured patients will die within 36 hours clearly indicates that a test of neurological integrity with more rapid acquisition and analysis, performed automatically and more frequently, would be of great diagnostic and prognostic value."

In the problem at hand, quantative studies and heuristic insight has led to a definition of four ordered classes for the evoked potentials. One of the classes (Grade 4, which indicates complete cortical failure) is very easily distinguished by the peak-detection program used to pre-process the waveform data. Differentiation of the other grades is an important problem, not least because resources are scarce and must be devoted to cases for which there is some hope of recovery.

A pseudo-algorithm has been constructed, but only to explain the practice of trained clinicians. In any case, the algorithm is not so well-specified that it can be implemented directly - features to be identified are peaks and troughs in a waveform and the inter-dependent timings are highly variable.

The subjective nature of the classification is illustrated well by a report that "the rate of agreement between two trials of expert grading ... was 77.3 percent". Ironically, a feed-forward neural network trained not on the raw signals but rather on a representation in terms of waveform peaks achieved a 77 percent accuracy.

Benchmarking neural network solutions against statistical techniques has been undertaken and the results seem favourable. A paper of interest here is [4], in which probability-based systems of analysis are held to present difficulties in the diagnosis of back pain because the assumption of statistical independence does not hold. As such, the neural network solutions need be evaluated only against the performance of clinical experts:

> "The best MLP and RBF networks outperform all three groups of clinicians on this problem. In the case of Spinal Pathology, the networks are much better that the clinicians and from a clinical point of view, this is the diagnostic class which it is most important to get right."

> "KNN, CCM and CCAD techniques seem better than the networks at diagnosing simple low back pain ... but are worse at diagnosing the other three types."

Learning was successful enough that "clinicians made the same misdiagnosis as the network".

Care is taken throughout that paper not only to establish the optimal number of hidden nodes for the data set, but also to choose desired output states which minimise the possibility of confusing the classes of disorder.

Scepticism about choice of architecture is illustrated well in [5], which develops a neural network capable of classifying muscle activity in an attempt to distinguish various forms of muscular distrophy. A wide range of feed-forward architectures and back-propagation learning algorithms are investigated, with a seemingly interminable analysis of learning curves, etc. Not all cases investigated produced satisfactory results in reasonable training time, showing that the application of neural network methods remains somewhat ad hoc.

Neural Networks is still a young discipline, with neither the body of experience nor the theoretical grounding which enable conventional techniques to be applied with reasonably little effort and fairly certain success. The conclusion in [5] that ANNs offer significant diagnostic yield (despite the trouble involved in constructing the network and verifying its performance) indicates clearly the existence of situations in medical physics where neural network solutions will play an important role, provided that sufficient care is taken over their construction.

# References

[1]    I.F. Croall and J.P. Mason, editors, "Applications of Neural Networks for Industry in Europe", United Kingdom Atomic Energy Authority (1991)

[2]    J.N. de Roach, "Neural Networks - An Artificial Intelligence Approach to the Analysis of Clinical Data", Australasian Physical & Engineering Sciences in Medicine, 12 100-106 (1989)

[3]    R.M. Holdaway, M.W. White and A. Marmarou, "Classification of Somatosensory Evoked Potentials Recorded from Patients with Severe Head Injuries", IEEE Engineering in Medicine and Biology, 9 43-49 (1990)

[4]    D.G. Bounds and P.J. Lloyd, "A Comparison of Neural Network and Other Pattern Recognition Approaches to the Diagnosis of Low Back Disorders", Neural Networks, 3 583-591 (1990)

[5]    C.N. Schizas, C.S. Pattichis, I.S. Schofield, P.R. Fawcett and L.T. Middleton, "Artifical Neural Nets in Computer-Aided Macro Motor Unit Potential Classification", IEEE Engineering in Medicine and Biology, 9 31-38 (1990)

# Data Representation And Generalisation In An Application Of a Feed-Forward Neural Net

D.L.Toulson

J.F.Boyce

C.Hinton

## 1  Introduction

This paper examines the role of a feed-forward neural network in an application involving the segmentation of medical images. Particular emphasis is focussed on the the choice of input to the network in terms of both its representation and content. It is shown that pre-processing of the input information by a statistical classifier leads to significant improvement in the networks performance. An examination is also made of the networks ability to generalise. In particular the importance of the use of validation data during training is demonstrated.

Due to the large size of the data sets used in this work it was found necessary to utilise the increased processing power of a transputer surface for the purpose of network training. The implementation of the *batch* back-propagation algorithm onto a MEIKO transputing surface is described along with an analysis of its improved performance over corresponding serial implementations.

Section 2 describes the application studied and gives an indication of the simplest way in which a network may be utilised in the segmentation process. Sections 3 and 4 develop the statistical classifier and describe how it is used to preprocess the input to the network. Section 5 discusses the implementation of network training onto the MEIKO transputer surface.

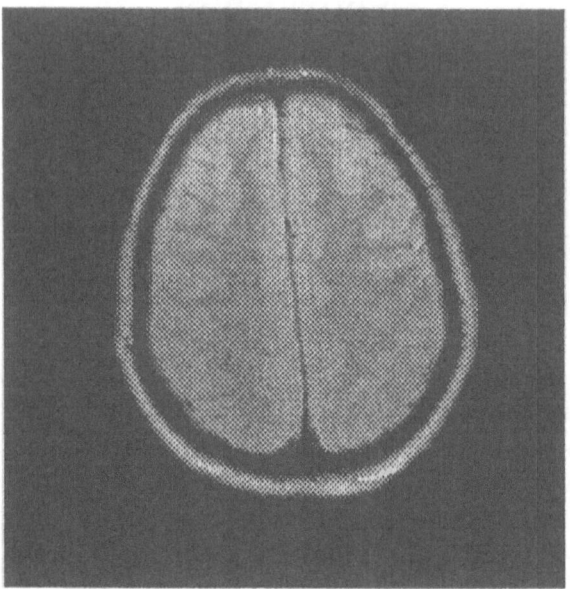

Figure 1: A typical MR image slice

Section 6 presents the results and comments on the relative efficiencies of the various representational schemes.

## 2   MR Image Segmentation

The application studied in this paper is the automated segmentation of multiple-slice Magnetic Resonance (MR) images of the head. A typical image slice is shown in Figure 1.

The purpose of the segmentation is to assign to the pixels of each image slice a representative anatomical label. A typical set of labels might be {Background, Scalp, Skull, Grey Matter, White Matter }. One purpose of such a segmentation would be to facilitate the automatic quantitative extraction of volumetric clinical data useful for instance to a neuro-radiologist planning radiotherapy. It is also a pre-requisite for the surface rendering algorithms used to allow the three dimensional display of two dimensional data sets.

In order to automate the segmentation process it is important to incorporate the type of knowledge a neuro-radiologist would use in performing the

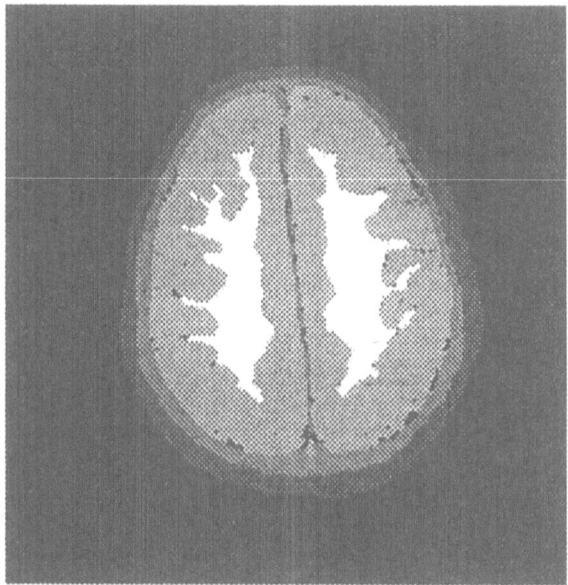

Figure 2: Manual Segmentation of Figure 1.

segmentation. To ensure this, a set of manually segmented images created by the majority decision of three neuro-radiologists was obtained. An example of a manual segmentation is shown in Figure 2.

If the segmentation process is to be performed on a pixel by pixel basis using grey level information as the labelling criteria it is already apparent how a neural network could be used. The manual segmentations along with the original grey level images could be used to obtain the input/output pairs necessary to train a suitable chosen network architecture. However the results shown in section 6 indicate that this somewhat *naive* use of a neural net produces poor results.

In the next section we develop a statistical classifier that acts as a pre-processing step on the grey level images. The results show that, by using the output of this statistical classifier as the net's input, rather than the raw grey levels leads to greater classification accuracy.

Neural Implementation

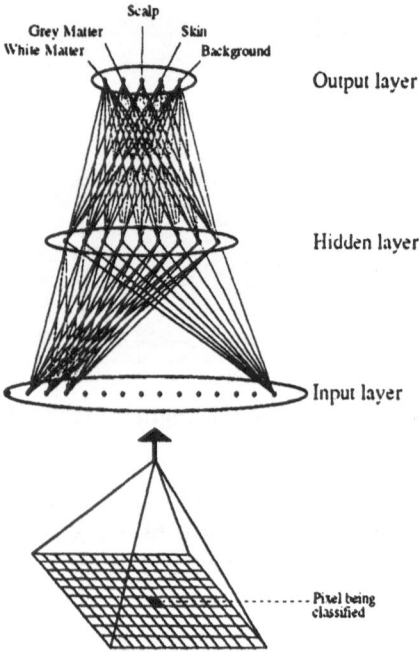

Figure 3: Probabilistic representation to net

# 3 Probabilistic Segmentation

The purpose of the statistical classifier will be to map the grey levels of the pixels of the original image onto an n-tuple of probabilities representing the chances that a pixel will belong to each of the n possible labels. The probabilities will be derived from the second order grey level statistics of the image obtained from the co-occurrence matrix [1]. The co-occurrence matrix has been shown to be a suitable feature space for this type of segmentation [2] and has the advantage of separating region and edge distributions allowing an 'interior-region only' analysis of the image statistics.

Consider an $N \times N$ digital image P. Let the pixels of the image be denoted by vectors $\vec{x} = (x, y)$, $x, y = 1...N$ and the intensity of a pixel be denoted by $i(\vec{x})$. Now assume the intensities of the pixels with a given label may be represented by a Random Variable. The distribution of intensities across the entire image may then be written as a finite mixture of the label conditional distributions,

$$p(i) = \sum_{a=1}^{A} \pi_a \, p_a(i), \qquad (1)$$

where $A$ is the number of distinct labels in the image, $\pi_a$ are the a priori probabilities of a pixel having label $a$ and $p_a(i)$ are the label conditional intensity distributions.

This result can be extended to the case where instead of single pixels, the joint intensity distributions of pairs of pixels are considered,

$$p_{\vec{\Delta}}(i, j) = \sum_{a,b=1}^{A} \pi_{\vec{\Delta};a;b} p_{\vec{\Delta};a;b}(i, j), \qquad (2)$$

where $\pi_{\vec{\Delta};a;b}$ is the joint probability that a pair of pixels separated by the vector $\vec{\Delta} = (\Delta x, \Delta y)$ have labels a and b, respectively and $p_{\vec{\Delta};a;b}(i, j)$ is the joint label conditional intensity distribution.

It is possible to split the $A^2$ distributions of the above equation into interior $(a = b)$ and boundary $(a \neq b)$ distributions,

$$p_{\vec{\Delta}}(i, j) = \sum_{a=1}^{A} \pi_{\vec{\Delta};a;a} p_{\vec{\Delta};a;a}(i, j)$$
$$+ \sum_{a \neq b} \pi_{\vec{\Delta};a;b} p_{\vec{\Delta};a;b}(i, j). \qquad (3)$$

The aim of this derivation is to find the probability of a pair of pixels belonging to a class $a$ based on the joint observation of their intensities $p_{\tilde{\Delta}}(a \mid i, a \mid j)$. Using Bayes Theorem we can write

$$p_{\tilde{\Delta}}(a \mid i, a \mid j) = \frac{\pi_{\tilde{\Delta};a;a} p_{\tilde{\Delta};a;a}(i,j)}{p(i,j)}. \tag{4}$$

The denominator of the above is a class independent normalising term that need not be explicitly calculated. This leaves the problem of finding the densities $\pi_{\tilde{\Delta};a;a}$ and $p_{\tilde{\Delta};a;a}(i,j)$. It has been shown [3] that these distributions may be estimated from the leading diagonal of the co-occurrence matrix. In particular, assuming that the distributions have a Gaussian form it is possible to write a Likelihood Function of the Gaussian parameters in terms of the co-occurrence matrix $S_{\tilde{\Delta}}(i,j)$,

$$\mathcal{L}_0(\psi) = \prod_{i,j=1}^{L} \left[ \sum_{a=1}^{A} \pi_{\tilde{\Delta};a;a} p_{\tilde{\Delta};a;a}(i,j) \right]^{S_{\tilde{\Delta}}(i,j)}, \tag{5}$$

where

$$p_{\tilde{\Delta};a;a}(i,j) = \frac{1}{2\pi |\Sigma_a|^{\frac{1}{2}}} \exp^{-\frac{1}{2}(\vec{z}-\vec{\mu_a})^T \Sigma_a (\vec{z}-\vec{\mu_a})}. \tag{6}$$

We wish to find the set of parameters $\psi$ which maximises $\mathcal{L}_0(\psi)$. This can be done by locating the stationary point

$$\frac{\partial \mathcal{L}_0(\psi)}{\partial \psi} = 0. \tag{7}$$

Unfortunately there exists no explicit solution to the above and moreover gradient descent techniques suffer due to the existence of singularities in $\mathcal{L}_0(\psi)$ [6]. A common technique used in overcoming this difficulty for the case of finite mixture distributions is the use of the EM (Expectation Maximisation) algorithm [4].

The EM algorithm operates by starting with an initial *guess* of the parameters and then iteratively updating them based on the actual sample data. For the case of a finite mixture of multivariate normals we can write the update equations in terms of the co-occurrence matrix,

$$\pi_{\tilde{\Delta};a;a} = n^{-1} \sum_{i,j}^{L} \omega_{i;j;a} S_{\tilde{\Delta}}(i,j),$$

$$\mu_{\tilde{\Delta};a;a} = n_a^{-1} \sum_{i,j}^{L} \omega_{i;j;a} S_{\tilde{\Delta}}(i,j)\vec{x}, \ ,$$

$$\Sigma_{\tilde{\Delta};a;a} = n_a^{-1} \sum_{i,j}^{L} \omega_{i;j;a} S_{\tilde{\Delta}}(i,j)(\vec{x} - \vec{\mu_a})(\vec{x} - \vec{\mu_a})^T,$$

where

$$n = \sum_{i,j}^{L} S_{\tilde{\Delta}}(i,j),$$

$$n_a = n\pi_{\tilde{\Delta};a;a},$$

$$\omega_{i,j,a} = \frac{\pi_{\tilde{\Delta};a;a} p_{\tilde{\Delta};a;a}(i,j)}{p(i,j),}$$

It is interesting to note the use of the *weights* $\omega_{i,j,a}$ in the above equations acting as fuzzy classifications of the data points. There are parallels between this algorithm and work done with neural networks. Perlovsky [5] for instance has used the EM algorithm as an integral part of his MLANS (Maximum Likelihood Adaptive Neural System) architecture realising the fuzzy classifications values as actual physically adjustable weighted links between processing elements.

# 4  Role Of The Neural Net

The statistical classifier developed in the previous section maps the original grey level information of the image into a set of fuzzy pixel classifications. One of the advantages of this type of representation over the 'raw' grey levels is that it allows an invariance to certain changes in the statistical nature of the grey levels of the input image common in image processing. For instance if the mean grey level of all pixels in the image were shifted by some amount or the amount of system noise were to increase then the adaptive parameters in the previous model should compensate for this, leaving the actual fuzzy classification invariant. For this reason the output of the statistical classifier was considered suitable as input to the neural network.

Pixels are classified one at a time using probabilistic labelling information from pixels located in a fixed neighbourhood around the pixel being classified (see Figure 3).

The final implementation decision to be taken about the network is how best to represent the probabilistic input data. In order to test their relative efficiencies three different input representations were considered,

- Simple - In the simple representation each label probability or grey level is assigned a single neuron. This is a costly representation in terms of input neurons. In the case of probabilistic input using a neighbourhood mask of size $M \times M$ with $A$ possible classes, $AM^2$ neurons are required.

- Single Neuron Coding - In this mode the label probabilities for a single pixel are combined to give a single input value,

$$input = \sum_{a=1}^{A} p(a) \frac{a-1}{A-1}. \tag{8}$$

A disadvantage of this scheme is that it will in general be degenerate although if the numbering of the labels is chosen such that the mean expected intensity of label $n+1$ is higher than that of label $n$ the effect of this degeneracy will be greatly reduced due to the nature of the statistical classifier.

- Interpolation Coding [8] - This coding is identical to the single neuron scheme except it incorporates a second neuron whose input is chosen such that the sum of the two inputs is always 1.00.

Using the manual segmentations, networks were trained using each of the above representations. During training use was made of a validation set This was composed of an additional set of input/output pairs appended to the usual training set. While training, the validation pairs are classified in the usual manner by the net and their errors calculated but the network does not perform the back-propagation stage. By monitoring the total error of classification of the validation set it is possible to monitor the generalisation ability of the net and training can be halted before the network begins *over-generalisation* occurs. Because of the size of the training sets used (training data taken from a single image using about 5 percent of the pixels and a $9 \times 9$ neighbourhood area occupies about 2.5 MB of memory) a version of 'batch' back-propagation [7] was implemented on a MEIKO transputing surface. The details of this implementation are given in the next section.

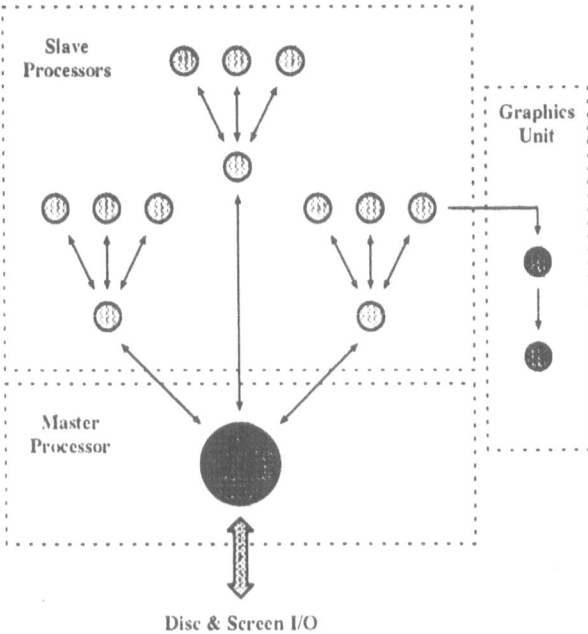

Figure 4: Meiko Process Distribution

# 5   Parallel Implementation

One of the most important decisions to be made when implementing an algorithm in parallel is that concerning the ultimate physical topology of the computing network and the distribution of processes within it. The approach taken for this problem was to make use of the, highly replicable, tertiary-tree structure (Figure 4) which has the highest level of interconnectivity possible for a tree structure, with a four link transputer. The inherent independence of processes, discussed later, suggests the use of a process farm distribution. Such a *farm* consists of a top level *master* sending out work to a number of *slave* processors, and consolidating the results of their efforts.

Clearly this structure requires a degree of data *routing* to allow communication between the master processor and the distant slave processors further down the tree. Furthermore this implementation offers a real-time graphical display of the network statistics during training, which is processed by a separate transputer - requiring further routing. Since the software is written entirely in OCCAM, the transputers native language, all routing is built into the program structure and optimised for this particular application.

Training the network is simply a matter of loading the full *training set* onto the networks master processor along with optional validation sets, up to the limit of the available memory. This large data set is then split into 12 equally sized blocks and distributed to the slave nodes of the network, along with the network profile and interconnection weight values. Each slave processor, contains an identical copy of the neural network with a unique training set.

The errors associated with each training vector are summed together for one iteration through the whole set, followed by a similar operation on the validation set. These results are passed back to the master processor, which sums the errors for all the slave nodes, returning the combined information to each one. Network weights for each slave node can now be changed using back-propagation, producing identical networks on every slave node.

The separate *graphics* unit allows analysis of the various errors associated with the training whilst the master processor provides for interactive adjustment of training parameters, extraction of graphs and storing of intermediate network weights, all independent of the network training.

## Performance

Using a reasonably small network training set, a single transputer running the complete system concurrently, manages to perform slightly faster in CPU time usage than a VAX 8700. With the addition of the full transputer network (12 slaves, 1 master/controller), there is an increase in speed over one transputer of approximately 12 times. This confirms the linear speed-up achievable with the addition of more transputers. Furthermore, relative communication overheads fall considerably as the data set is increased in size, due to the local storing of training data by each slave processor.

## 6  Results

Figure 5 summarises the classification ability of networks trained with a variety of input representations. The results clearly illustrate the deficiency of

| Input Format | Represen-tation | Mask Size | Input Layer | % Accuracy |
|---|---|---|---|---|
| Grey Level | Simple | 17x17 | 289 | 84.2 |
| Probabil-ities | Single Neuron | 17x17 | 289 | 85.4 |
| Probabil-ities | Interpol-ation | 9x9 | 162 | 87.5 |
| Probabil-ities | Interpol-ation | 13x13 | 338 | 89.1 |
| Probabil-ities | Simple | 7x7 | 245 | 91.3 |
| Probabil-ities | Simple | 9x9 | 405 | 92.4 |

Figure 5: Relative classification accuracies of the representational schemes

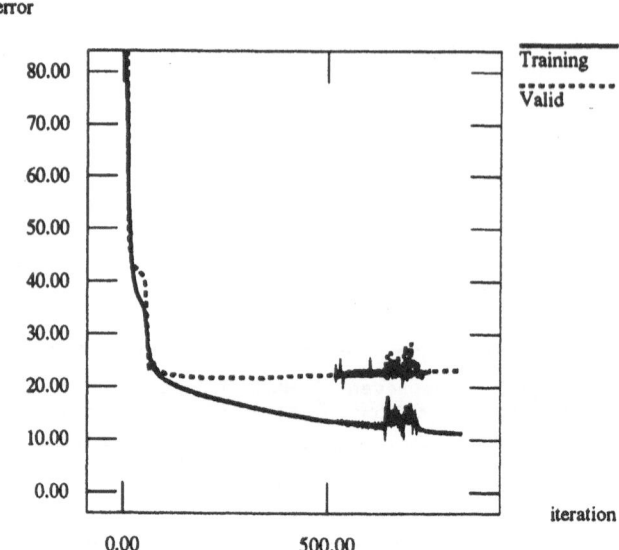

Figure 6: Training statistics illustrating the *over-generalisation* effect

using simple grey level input to the network. Of the representations using probabilistic label input it was found that the *simple* representation lead to the greatest classification accuracy. It is interesting to note however the increase in performance of the single neuron representation using interpolation coding over the compact single neuron representation.

Figure 6 shows the training statistics of the most successful representation (9x9 Simple). The figure shows the effect of over-generalisation during training, apparent after 1000 epochs. Although the training error is continuing to fall, the error in the validation set is clearly rising. To combat this, the criteria for stopping the training process was made to be the minimisation of the validation set error and not the training error.

Figures 7 and 8 illustrate the improvement in segmentation provided by the network over the statistical classifier alone. A particular improvement is seen in the classification of the skin/skull labels.

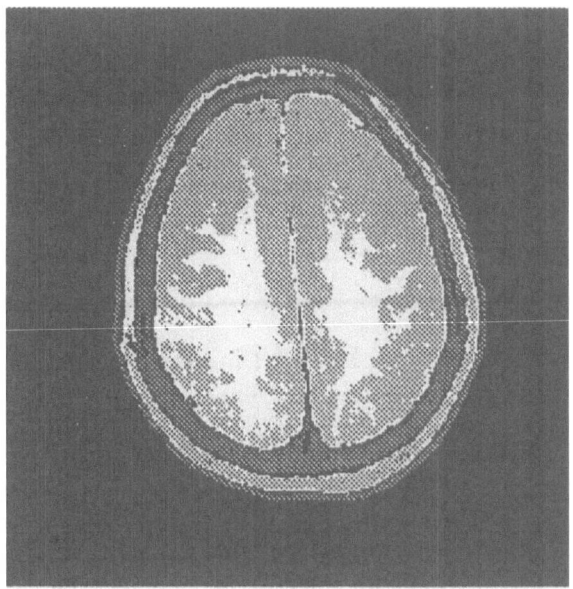

Figure 7: Hard classification of the probabilistic segmentation (85.6 % accuracy)

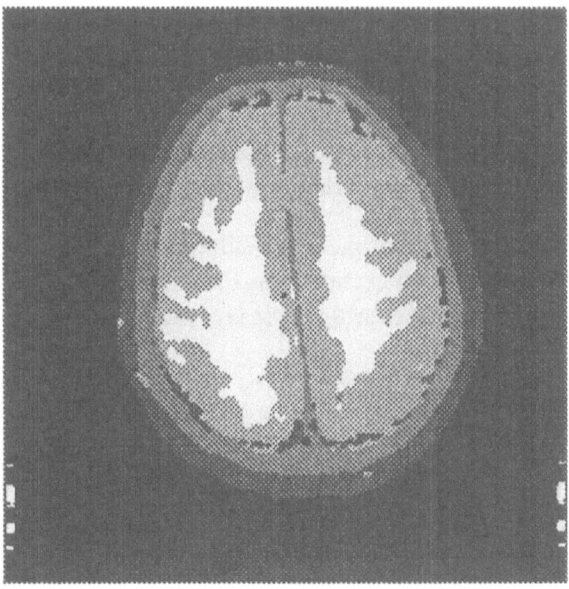

Figure 8: Neural classification using Figure 7 as input (92.4 % accuracy)

# Acknowledgement

Darren Toulson wishes to acknowledge IBM Scientific Center, Winchester for their support of his CASE studentship. The authors also wish to thank the Gregorio Marañon Hospital in Madrid for supplying the manually segmented images as part of the AIM project COVIRA.

# References

[1] R.M.Haralick, K.Shanmugan, Its'Hak Dinstein. Texture Features for Image Classification, IEEE Trans. Sys. Cyber., SMC-3 , 610-621. 1973.

[2] J.F.Haddon, J.F.Boyce Image Segmentation by Unifying Region and Boundary Information, PAMI-12, 929-948, 1990.

[3] D.L.Toulson, J.F.Boyce Segmentation Of MR Images Using Neural Nets, in P.Mowforth (ed.), BMVC91, 284-292, Springer-Verlag, 1991.

[4] A.P.Dempster, N.M.Laird, D.B.Rubin. Maximum Likelihood Estimation From Incomplete Data via the EM Algorithm (with discussion), J.R. Statis. Soc. B 39, 1-38. 1977.

[5] L.I.Perlovsky M.M.McManus Maximum Likelihood Neural Networks for Sensor Fusion and Adaptive Classification. Neural Networks 4-2 89-102. 1991.

[6] D.M.Titteringdon A.F.M.Smith U.E.Makov Statistical Analysis of Finite Mixture Distributions. New York:John Wiley & Sons. 1985.

[7] D.E.Rummelhart, G.E. Hinton, R.J.Williams Learning Internal Representations By Error Propagation. In Parallel Distributed Processing, Chapter 8. MIT Press, Cambridge, Mass. 1986

[8] D.H.Ballard Interpolation Coding: A Representation For Numbers In Neural Models. Biological Cybernetics 57 389-402 1987.

# Simulated Ultrasound Tomographic Imaging of Defects

D.M. Anthony

E.L. Hines

D.A. Hutchins

J.T. Mottram

Department of Engineering

Warwick University

Coventry CV4 7AL

England

April 13, 1992

### Abstract

Simulations of ultrasound tomography have demonstrated that artificial neural networks can solve the inverse problem in ultrasound tomography. A highly simplified model of ultrasound propagation was constructed, taking no account of refraction or diffraction, and using only longitudinal wave time of flight (TOF). TOF data was used as the network inputs, and the target outputs were the expected pixel maps, showing defects (grey scale coded) according to the velocity of the wave in the defect.

The effects of varying resolution and defect velocity were explored. It was found that defects could be imaged using time of flight of ultrasonic rays. Further simulations also investigated a configuration that could be implemented experimentally. It was found that in this case the resolution had to be limited to achieve successful reconstructions.

## 1   Introduction

Ultrasonic waves have been used in many applications to determine the properties of a material or structure, and there are many different ways of analyzing the data and providing images. Although conventional methods of image reconstruction exist, there may be cases where they do not perform effectively, and in such cases a neural network approach may be of interest.

In this paper, we are particularly interested in the use of ultrasonic techniques for the detection of delamination cracks in multi-layered composite plates [1]. Such composite plates are fabricated by stacking together a number of plies of unidirectional fibre-reinforced polymer material. Each ply has directional properties with a high specific strength and specific stiffness in the direction of fibre alignment. To compensate for the much lower specific properties in directions other than that corresponding to the fibre alignment, it is standard practice to build up structural laminated plates with the plies oriented at $0°$, $90°$ and $+/-45°$ to the principal loading direction. For aerospace applications [2], the fibre reinforcement is usually carbon and a typical laminate may consist of 40 (5 mm thick) to 120 layers (15 mm thick).

One problem with the carbon fibre reinforced polymer plates is that they are susceptible to the formation of delamination cracks. These cracks are formed between the fibre layers in the resin rich interface. They are a consequence of the material possessing poor through thickness properties. The presence of delamination cracks can seriously limit the compressive load bearing capacity of a laminate. For this reason it is important that we have a non-destructive instrument capable of detecting the presence of delamination cracks in multi-layered plates. Delamination cracks may be formed during the manufacturing process or by the laminate incurring damage from impact with a foreign body. It has been concluded [1] that delamination cracks in the wings of aircraft may become a source of concern if their diameter exceeds 20 mm. During routine maintenance of existing aircraft with composite structural components, several slow non-destructive evaluation techniques are now used to see whether or not delaminations are present. To supersede these traditional detection methods, an ultrasonic scanning system is being developed which will use artificial neural networks (ANNs) to provide rapid analysis of data. The final scanning system should provide an instrument for the rapid detection of large delamination cracks in multi-layered composite plates. The purpose of the present paper is to describe simulations of such a system, to determine how the ANN approach performs in ultrasonic imaging applications.

One method for producing an image from ultrasonic measurements is tomographic reconstruction. This has been used, primarily in medical imaging, to provide cross-sectional images. In most cases, a set of transducers are scanned around the object, which is usually immersed in water, and waveforms along pre-selected paths are sampled to result in a transmission image [3, 4]. Applications in NDT have also been described [5]. In all cases, it is necessary to select the paths through the object carefully, so as to fulfil the requirements of the image reconstruction algorithm. There are many approaches that can be used to form the image from experimental data, although two of the most common are transform-based methods [6] and series expansion methods [7].

A complication that exists in ultrasonic tomography that is not met in x-ray imaging is that diffraction and refraction effects complicate the transmission of energy from source to receiver. Although attempts have been made to counteract such phenomena by modifying the reconstruction algorithm, the results are not always satisfactory. In the particular case of the testing of complex materials such as fibre-reinforced polymer composites, additional factors such as anisotropic elastic properties cause additional complications to arise. This was the principle reason for investigating the use of neural networks for the reconstruction of tomographic images.

ANNs are networks of simply interconnected processing units which can learn their behaviour according to inputs received during a training phase. ANNs are based loosely on the structure of interconnected neurons in the brain, but in most cases do not attempt to model the way the brain behaves. ANNs enable solutions to be found to problems where algorithmic methods are too computationally intensive or do not exist [8]. ANNs also offer significant speed advantages over conventional techniques. The problem of image formation therefore seemed to be a suitable application for ANNs. The investigations described in this paper have concentrated on the Multi-Layer Perceptron (MLP) network using the back propagation learning technique as described by [9]. This combination is the simplest and most widely used neural network [10], and is used here for simulations into the possible application of ANNs to tomographic reconstruction.

ANNs have been used [11], to recognise digits scanned by ultrasonic "eyes" and digits the network was not trained on were reconstructed with some success. ANNs have determined the radius of a cylindrical object, given the real and imaginary pressures from an array of 16 transducers surrounding the object [12]. ANNs may be used to process data, for example in medical ultrasound, delay noise has been reduced in simulations where an ANN pre-processed echo delays prior to beamforming [13]. Other NDT applications using ANNs include the inversion of eddy currents to give flaw size [14]. ANNs have also been used in other tomographic modalities, e.g. Laser Scattering Tomography [15], and by [16] to improve a tomographic image that had already been formed by conventional techniques. The aim of this paper is to report work which has been performed using simulation experiments to solve the inverse problem in ultrasound imaging. The final target is to produce tomographic maps of anisotropic composite materials containing defects which may be the result of service load conditions (e.g. carbon fibre, reinforced plastics). These are particularly difficult to image using ultrasound as the orientation of the fibres in the resin matrix affects the wave propagation, and the fibres act as channels for the waves, thus adding to the above mentioned diffraction and refraction effects. ANNs might be successful in solving this mapping problem.

Prior to attempting the more difficult anisotropic case, it was decided to initially determine whether isotropic media could be imaged. Problems in this type of material would be expected to be less severe than the anisotropic case, and thus the more difficult problem would be unlikely to be solved unless the simpler case was shown to be soluble. Simulations were undertaken as a prelude to experimental data acquisition, to allow tighter control and for speed of development.

## 2    The Simulation Model

A thin sheet specimen of square cross-section was created, which was assumed to have constant in-plane wave velocity in all directions (ie. isotropic). A defect lying in the plane, may in principle be of any size, however a finite imaging resolution will limit the representation or display of the defect. The resolution will be (say) $N$ by $N$. It was assumed that if the defect lies in any of the $N^2$ cells, the cell will be designated as having the defect velocity, and if a cell contains no part of a defect it will be designated as background material wave velocity. Thus where a defect lies across the boundary of two cells, both cells will be assumed to have the defect wave velocity. Figure 1 illustrates the scanning method that was adopted for the simulations. An ultrasonic transmitter was assumed to move around the outside of the pixel area, and the ultrasonic signal to be received at all other receiver locations not on the same edge as the transmitter.

(The transducers around the specimen may act as transmitters or receivers.)

In subsequent discussions the actual defect dimensions will be referred to as the "ideal" image. The image which results from a given output image resolution will be called the "target" image. The ANN will be given inputs which will be the TOF of ultrasound rays, and will be trained to give the associated target pixel values.

# 3   Time Of Flight Calculations

Initially a very simple model is proposed. The ultrasound wave propagation is assumed to travel in straight lines, with constant velocity in the medium, and with a lower constant velocity in the defects. No account is taken of diffraction, refraction, reflection, or the effects of anisotropy. In addition only longitudinal waves are assumed, and no account is taken of shear waves etc. Consider a specimen surrounded by a medium of constant propagation. The specimen can be split into cells within which a constant velocity may be assumed. The simplistic assumption is that the TOF may be computed by taking a straight line between the transmitter and any of the 12 receivers, and by working out the distance traveled in the cells [3]. For a given angle of incidence $\theta_i$. the TOF for ray $j$ at angle $i$, which is the measured value, is given by :-

$$T_{ij} = \sum_{k=1}^{N} \frac{L_{kj}^{(i)}}{C_k} + \frac{D - \sum_{k=1}^{N} L_{kj}^{(i)}}{C_m} \tag{1}$$

where $D$ is the total distance between transmitter and receiver, $L_{kj}^{(i)}$ is the length of the kth cell traversed by the jth ray at orientation i, $C_m$ is the velocity of wave propagation in the medium surrounding the specimen, $C_k$ the (constant) wave velocity in the kth cell, and $N$ is the number of cells. If one denotes $n_k = \frac{1}{C_k} - \frac{1}{C_m}$, one has a parameter to be reconstructed ($n_k$) subject to :

$$\sum_{k=1}^{N} L_{kj}^{(i)} n_k = \tau_{ij} \left\{ \begin{array}{l} i = 1, ...I \\ j = 1, ...M \end{array} \right\} \tag{2}$$

where $\tau_{ij} = T_{ij} - \frac{D}{C_m}$ This forms a set of $M$ equations, if $M$ rays are considered, in $N$ unknowns which may be solved using algebraic reconstruction techniques (ART), convolution methods, fourier analysis, or diffraction methods [3]. In this study a solution to the equations will be attempted using ANNs. The $T_{ij}$ will be given as inputs and the required image as targets to the ANN. In principle a linear network can solve this equation, as the set of equations are linear. Non linear nets will be employed since in practice non-linearities may be encountered, for the TOF calculations have built in assumptions which make the linear equations an approximation to reality only. Thus the non-linear net needs to be tested in the linear simulation because it will be used in studies based on experimental data.

The distances are calculated using simple geometry, where the transducer locations and the rectangular defect co-ordinates are known. In a reconstruction each transducer would in turn be the transmitter, and the remaining transducers on the other three sides would be receivers.

The net will be required to solve the inverse problem of equation ( 2), ie.

$$n_k = F(T_{ij}) \tag{3}$$

# 4  Experimental Procedures

Given an array of transducers, the effects of varying the resolution of the final image, and the ratio of defect wave velocity to background material wave velocity were explored. The imaging of rectangular single defects of arbitrary dimensions was attempted.

A set of 16 transducers were assumed to be arranged spaced equally around a material of square cross-section. TOF paths were computed from all transmitters to every other receiver not on the same edge. This gave 192 values. There was redundancy in the data as each path was given twice, due to the symmetry of the problem, but as in real simulations, especially in anisotropic media, these TOF would not be identical due to noise and other effects, it was therefore decided to include all values. The net with all inputs used is potentially more liable to fall into uninteresting one to one mappings (ie. act as a look-up table) than one using half the inputs, and thus one needs to test the net for this potential problem, if one is later to use all inputs in an experimental study. These were fed into an input layer of 192 units, which were fed into a hidden layer of 48 units, and the hidden layer was connected to an output layer which gave pixel values for a reconstructed image. The figure of 48 hidden units is somewhat arbitrary; unfortunately there is little theoretical justification for using any particular architecture. Too large a number of units in the network will give slow convergence, and may not generalise well; too few units will not allow a solution to the problem. The architecture used in this study gave convergence, and the nets showed similar performance on test data as training data, thus the nets did generalise. The nets also converged in a reasonable time.

In these simulations a fan-beam geometry is assumed (see Figure 1). The equivalent setup in parallel beam tomography would be to have $K$, the number of projections, as given by :-

$$K = (N_d/4)(3N_d/4), \qquad (4)$$

where $N_d$ is the total number of receivers, as each transmitter sends rays to the receivers on all other three sides. It is known in such a system that to avoid aliasing [17], the number of projections must be given by :-

$$K \geq \frac{\pi}{2}N_d. \qquad (5)$$

Thus the minimum number of receivers that will avoid aliasing is given by :-

$$3N_d^2/16 \geq \frac{\pi}{2}N_d \qquad (6)$$

or :-

$$N_d \geq \frac{8\pi}{3} \qquad (7)$$

In [16] it was found that recognition of simulated "lungs" dropped to very low levels when the ratio of receivers to projections fell below the minimum specified by equation ( 5), though the error level of the output image was able to be reduced substantially by the network. The figure of 12 receivers used in this study thus should not suffer from the problem of aliasing.

In all simulations error back propagation was employed [18]. The network parameters of learning rate ($\eta$) and momentum ($\alpha$) need to be given values. The lower the value of $\eta$ the more likely convergence is achieved, but very low values may slow convergence unnecessarily. An $\alpha$ close to unity speeds convergence, and [18] suggest a high $\alpha$ of about 0.9 with a low $\eta$ is superior to a higher $\eta$ with no momentum term. An $\eta$ of 0.001 and $\alpha$ of 0.9 were used, and these values were found to allow convergence in all cases.

# 5 Results of Simulations

## 5.1 Defect Types

Arbitrary dimensioned rectangular defects were randomly created for the training set. The resolution of the target image was set, and if a pixel contained any of a defect, that pixel was allocated to the defect, otherwise it was allocated to the background material. Pixels were given values according to the wave velocity of defect or background. Thus images were over-represented in size as defects were rounded up in size for the target pixel map, and a grey scale was produced whereby white indicates the highest velocity, and black zero velocity. Initially 4 by 4 resolution images were created using a defect velocity 10% that of the background velocity. This large reduction in velocity value due to the presence of a defect may not be realistic, but a large difference in velocities was found to allow easier network learning.

As Figure 2 shows, the ANN restored image locates the defect with some success.

## 5.2 Increasing Resolution

Increasing the resolution of the output pixel map to 10 by 10 gives better restored images than the 4 by 4 case, see Figure 2. (nb. The number of the receivers remains at 4 by 4.) Defects whose length to width ratio is high are not detected in either resolution. This is likely to be due to the fact that the defect effects the TOF between transmitter and receiver very little, and the ANN has insufficient variability of its inputs to reconstruct the defect.

The ability of a network to converge is often shown by plotting total sum square error ($tsse$) against number of epochs of training, which is defined as :-

$$tsse = \sum_{p=1}^{N^p} \left( \sum_{i=1}^{N^o} (t_i^p - o_i^p)^2 \right) \tag{8}$$

where $N^p$ is the number of data patterns and $N^o$ the number of outputs, $t_i^p$ and $o_i^p$ are the target and output values respectively for the $p$th pattern and $i$th target or output value.

Figure 3 shows $tsse$ per pixel of networks with various resolutions, which shows the trend towards lower error as the resolution (number of pixels in the reconstructed image) increases.

## 5.3  Repeatability

To test whether the ANNs were giving consistent and repeatable results, several runs were made using different initial random weights. As the ANNs were found to converge within two to five epochs (see Figure 3) the nets were trained for 10 epochs each. The nets showed very similar behaviour on repetition. The *tsse* in every case converged within 2-3 epochs, and was similar for both training and test (ie. using data that the net had not been trained on) data. The absolute value of weights from input to hidden, and hidden to output nodes slowly increased with time.

## 5.4  Amendments to Back Propagation Algorithm

To improve images two changes were made to the training algorithm. The majority bias problem occurs when most of the outputs of the net are set to the same value most of the time, and the net can reduce the *tsse* quickly by fixing the outputs at this value, ie. the network gets "stuck" at these outputs. As [19] have shown, the use of thresholding may improve images where the majority bias problem is apparent. In the case of small defects, most of the outputs are set to the background material, and thus this may partly explain the poor performance of the network to detect small defects. Thresholding was applied to the network, whereby any output which was within 10% of its target contributed no error to the back propagation algorithm. This would be expected to allow the network to concentrate on the smaller number of defect pixel outputs. However the performance was not improved in 10 repeated runs using this technique where the criterion of good performance was receiver operating characteristic curves (ROC, see section  5.5).

To speed convergence of nets, dynamic parameter tuning may be employed, as in [20]. In this technique the learning rate and momentum are altered according to whether the *tsse* is decreasing or increasing. Where the *tsse* decreases from one epoch to the next, $\eta$ is increased by a scalar (we used 1.1), and where the *tsse* increases $\alpha$ is set to zero, and $\eta$ reduced by a scalar value (we used 0.7). In 10 repeated runs little advantage was found using this method in terms of speed of convergence or the performance of the net. In every case $\eta$ increased for the first few epochs (2-5 typically) and then reduced.

## 5.5  Ratio of Defect to Background Wave Velocity

Reducing the difference between background and defect velocities decreases the *tsse* for single defects (Figure  3). However looking at the reconstructed images, the ANN does not perform any better, and subjectively appears to be worse, see Figure 4. It might be the case that the ANN reduces the *tsse* by setting the output pixels to a value midway between defect and background. The different slope of the two cases indicate that they may be learning different mappings. To test the relative performance of the nets Receiver Operating Characteristic (ROC) curves were constructed. This technique is well known in medical informatics, and for a description see [21, 22]. Essentially a graph is plotted of the true positive rate against false positive rate for a series of confidence levels. This technique has been applied to ANN analysis [23, 24]. An area significantly above 0.5 indicates that the network is not giving random classifications.

Table 1: Area under ROC curve for various defect velocities

| Defect/Background Velocity | Mean Area | Standard Deviation |
|---|---|---|
| 0.1 | 0.639 | 0.009 |
| 0.5 | 0.634 | 0.009 |
| 0.9 | 0.562 | 0.018 |

The area under the ROC curves for 10 repeated runs was calculated for the defect velocity of 50% and of 10% that of the background velocity. Using the t test of means, there was no significant difference between the two sets ($p > 0.1$). To test further the effect of defect velocity, a training set was constructed which had defect velocity 90% that of background. There was a significant difference between this set and, separately, both the other two sets (10% and 50%, $p < 0.0001$), see Table 1. For 10% versus 50% the variances were not significantly different. For both 10% and 50% versus 90% defect velocity, there was a difference in the variances, and this was taken account of in the t test using Satterthwaite's approximation for degrees of freedom [25] (The SAS package was used to perform the tests).

# 6    Extension to a Real Testing Geometry

Further simulations were performed to determine whether the ANN approach could be used to detect the location and size of a single defect, within a defined square pixel area. An experimental set-up was simulated, where a moveable source was taken around three sides of the square pixel area, with the fourth side containing a fixed set of detectors. The arrangement would thus be as shown in Figure 1, but now the upper side contains receivers only, with the source at the locations shown on the other three sides. This geometry was chosen to facilitate scanning, as in practice four fixed receivers could be connected permanently to four channels of amplification and digitization, leaving a single moveable transducer to be scanned around the object to the source positions in turn.

The simulations were also intended to illustrate how a given ANN could be trained on experimental data. This would be most easily achieved experimentally by moving a single defect through the area of interest, and recording a data set with the defect in each pixel location. With the ANN trained on such data, the task would be to reconstruct the position of such a defect, if it was present, from a data set with an unknown defect location. Simulations were performed initially as above, using different pixel densities, and with the single pixel defect having various values in the range 10%-50% of the background velocity. The results showed that the ANN did not perform in a satisfactory manner. This is probably due to the majority bias problem, encountered in the simulations described earlier. Even in a 4 x 4 pixel problem, where the defect pixel occupies some 6% of the image area, the network failed to distinguish between a defect-free image and that with a defect.

To investigate this further, the single defect was surrounded by nearest-neighbour pixels, with a value intermediate between the defect velocity and background. This effectively smeared the defect signal over a larger area. Even in this case, the network did not perform well. This gives further evidence that the ANN could not be trained properly on single pixel defects. It was thus decided to train the networks as before, namely to use a training set comprised of defects with a wide variation in defect size, shape and location, and then to attempt to reconstruct the location of a single pixel defect from a test set. Using a 4 x 4 pixel array, the results are as shown in Figure 5. Note that the defects have been located in the correct position, with an amplitude which is not too far from the target value. Further simulations with a higher pixel density and a corresponding increase in the number of source and receiver positions have been performed. The results indicate that the location success rate of a single pixel defect decreases. This is because a single pixel defect occupies a smaller proportion of the whole image as the pixel density increases, a phenomenon noted in the earlier simulations with sources and receivers on all sides of the square pixel area. The authors are currently investigating methods for improving the detection rate.

# 7    Conclusion

A simplistic ultrasound simulation has been constructed. An ANN solved the inverse problem, ie. given the TOF data it was able to construct a pixel image.

Various conclusions may be drawn from this study :-

- The network behaviour was repeatable.

- As the resolution of single arbitrary dimensioned rectangular defects increases, the *tsse* per pixel decreases. This may be due to a reduction in rounding up of defects for the targets.

- The defect velocity is critical. If the defect velocity is close to the background velocity, the defect recognition rate reduces. This may have importance when analysing experimental data. If the data shows small differences for TOF comparing defect and non defect paths, one may need to pre-process the data to increase the dynamic range.

- The use of thresholding errors from the output layer did not solve the majority bias problem.

- Dynamic learning did not improve the convergence in these simulations. target pixels

- Simulations of an actual experimental arrangement showed that a 4x4 pixel grid might be reconstructed, but further refinements would be necessary to result ina higher resolution image. ,

Using ANN techniques may provide alternative methods of reconstructing tomographic images. It may be the case that conventional tomographic reconstruction will continue to be used in isotropic materials, and that ANNs may be used in anisotropic materials. Another possibility would be the marriage of the two techniques, such that the conventional method produces an output which is improved by a neural net (e.g.).

Future work planned by the authors will explore alternative network designs on the simulation data (e.g. neural trees [26]). An experimental tomographic setup has been constructed, which will allow data to be collected on specimens with known defects. The data will be analysed to determine the most appropriate transformation of the data for input to the neural net. Initial study indicates the ratio of fourier transform amplitudes, or the area under the fourier transform, may allow a single number from each ultrasonic wave to be used. These parameters seem to be affected when a defect lies in the path of the ultrasonic wave. If experimental data is reconstructed with some success, the trained neural network could be implemented in hardware to form an imaging system.

## Acknowledgments

This project is funded by the Science and Engineering Research Council (S.E.R.C.), U.K.

# References

[1] Stone D.E.W. and Clarke B. Non-destructive evaluation of composites - an overview. In Matthews F.L., Buskell N.C.R., Hodgkinson J.M., and Morton J., editors, *6th International Conference on Composite Materials and 2nd European Conference on Composite Materials*, volume 1, pages 2–59, 1987.

[2] Middleton D.H. ed. *Composites Materials in Aircraft Structures*. Longman Scientific Technical, 1990.

[3] Mueller R.K, Kaveh M, and Wade G. Reconstructive tomography and applications to ultrasonics. *IEEE Proceedings*, 67:4:567–587, 1979.

[4] Greenleaf J.F. and Bahn R.C. Clinical imaging with transmissive ultrasonic computerized tomography. *IEEE Trans. Biomed. Eng*, 28:177, 1981.

[5] Eberhard J.W. Ultrasonic tomography for nondestructive evaluation. *Ann. Rev. Mater Sci*, 12, 1, 1982.

[6] Kak A.C. and Slaney M. Principles of computerized tomographic imaging. *IEEE Press*, 1988.

[7] Censor Y. Finite series expansion reconstruction techniques. *Proc. IEEE*, 71:409–419, 1984.

[8] Lippmann R.P. An introduction to computing with neural nets. *IEEE ASSP Magazine*, 4:2:4–22, 1987.

[9] Rumelhart D.E and Hinton G.E and Williams N. Learning internal representations by error propagation (in parallel distributed processing ed rumelhart and mclelland), 1986.

[10] Hecht-Nielson R. Neurocomputing : Picking the human brain. *IEEE Spectrum*, MARCH:36–41. 1988.

[11] Watanabe S and Yoneyama M. Ultrasonic robot eyes using neural networks. *IEEE Transactions on Ultrasonics, Ferroelectrics and Frequency Control*, 37(3 May):141–147, 1990.

[12] Conrath B.C, Daft C.M.W, and O'Brien W.D. Applications of neural networks to ultrasound tomography. *Ultrasonics Symposium*, pages 1007–1010, 1989.

[13] Nikoonahad M and Liu D.C. Medical ultrasound imaging using neural networks. *Electronic Letters*, 26(8 Apr 14):545–546, 1990.

[14] Mann J.M, Schmerr L.W, and Moulder J.C. Neural network inversion of uniform-field eddy current data. *Materials Evaluation*, Jan:34–39, 1991.

[15] Gonda T., Kakiuchi H, and Moriya K. In situ observation of internal structures in growing ice crystals by laser scattering tomography. *Journal of Crystal Growth*, 102(1-2 Mar-Apr):179–184, 1989.

[16] Obellianne C, Fogelman Soulie F, and Galibourg G. Connectionist models for image processing. In Simon J.C, editor, *From pixels to features, COST13 workshop*, pages 185–196. Elsevier, 1989.

[17] Platzer H. Optical image processing. In *Proc. 2nd Scandinavian conf. on image analysis*, pages 128–139, 1981.

[18] Rumelhart D.E and Hinton G.E and Williams N. Learning representations by back-propagating errors. *Nature*, 323 NO 9:533–536, 1986.

[19] Chiu W.C. Anthony D.M. Hines E.L. Forno C. Hunt R. And Oldfield S. Selection of the optimal mlp net for photogrammetric target processing. In *Iasted Conf. Artificial Intelligence App. & Neural Networks*. pages 180–183, 1990.

[20] Vogl T.P., Mangis J.K., Rigler A.K., and Zink W.T. and Alkon D.L. Accelerating the convergence of the back-propagation method. *Biological Cybernetics*, 59:257–263, 1988.

[21] Todd-Pokropek A.E. The comparison of a black and white and a color display : An example of the use of receiver operating characteristic curves. *IEEE Transactions*, MI-2:19–23, 1983.

[22] Swets J.A. Roc analysis applied to the evaluation of medical imaging techniques. *Investigative Radiology*, 14 PT 2:109–121, 1979.

[23] Meistrell M.L. *Evaluation of neural network performance by receiver operating characteristic (ROC) analysis: examples from the biotechnology domain*, pages 73–80. Elsevier, 1990.

[24] Anthony D.M. *The Use of Artificial Neural Networks in Classifying Lung Scintigrams*. PhD thesis, University of Warwick, 1991.

[25] Sas users's guide, statistics.

[26] Sankar A and Mammone R.J. Optimal pruning of neural tree networks for improved generalization. In *Proc. 5th International Joint Conference on Neural Networks*, Seattle, 1991. IEEE.

Figure 1: Fan ray tomogram as used for ANN simulation (left). The material is assumed to be split into NxN cells. Within each cell, the wave velocity is assumed constant, (here 100 cells are shown). The ultrasonic beam is sent from a transmitter (marked as closed circles around the perimeter of the square material), and are received by transducers (open circles) along the ray path indicated by arrows. The transmitter is then moved to each pixel location in turn. Transducers setup for real testing tomography (right). Transmitters (closed circles) send ultrasonic rays to 4 fixed receivers only (open circles)

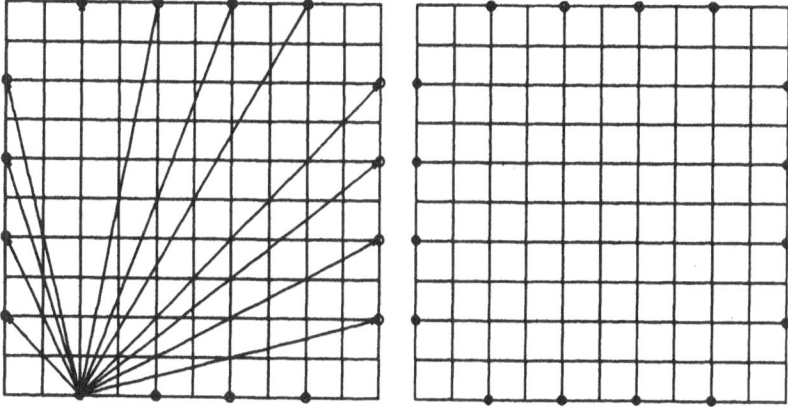

Figure 2: Target (left) restored (centre) and ideal (right) images of randomly created (single) rectangles. The ideal images are the actual dimensions of the simulated defect. The target image is created by using a given pixel resolution (top 2 images 4x4, bottom 3 images 10x10) so the defect is rounded up in size, this is the image the ANN is given during training as a target. The ANN after training gives the centre image when presented with the TOF data.

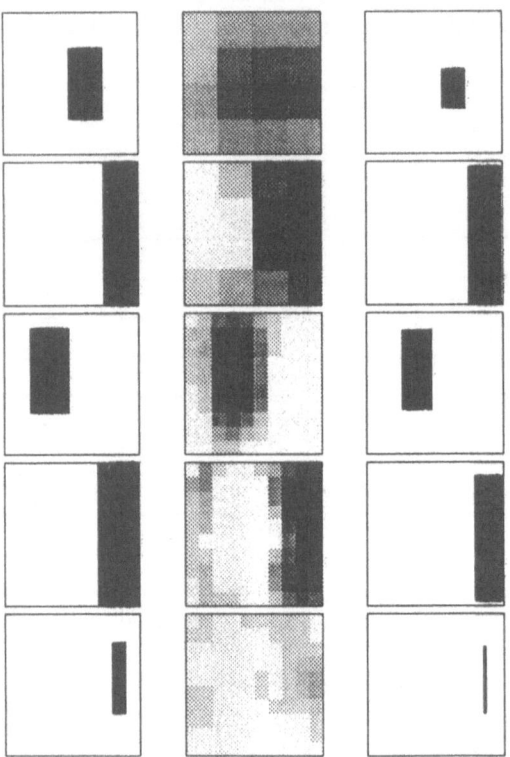

Figure 3: *tsse* per pixel for various resolutions against number of epochs of training - single defect (top graph). As the resolution increases the *tsse* drops. Key: _ 2 by 2, -.-. 3 by 3, ... 10 by 10 pixels. Lower graph - *tsse* per Pixel for 4 by 4 Resolution against epochs of training, Single Defect for 2 different defect velocities. The *tsse* for the defect velocity 50% that of background material is lower than the *tsse* of a defect with wave velocity 10% that of background material (lower graph). Key: _ 10 % defect velocity, -.-. 50 % defect velocity

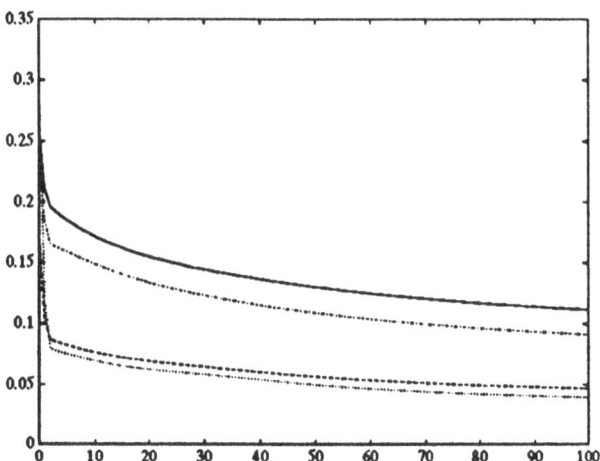

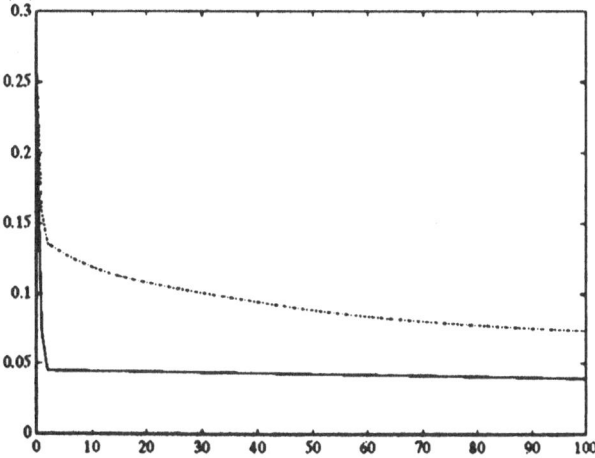

Figure 4: Subjectively the image with the lower *tsse* (50% defect velocity) was not improved. Top figures show target (left) and output (right) from defect velocity 10% that of background material. Bottom figures show target (left) and output (right) for 50% defect velocity

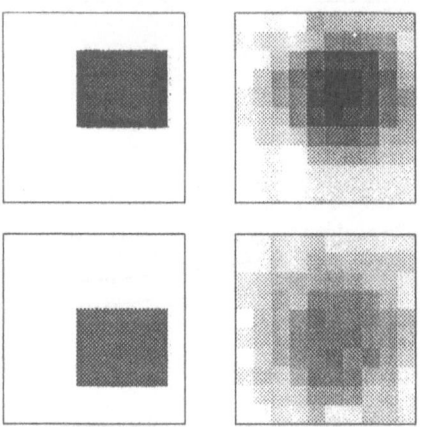

Figure 5: Targets and outputs of net trained on arbitrarily sized rectangles, given test set of pixel sized defects in 4x4 resolution. Each target has the output actually given by the net to its left.

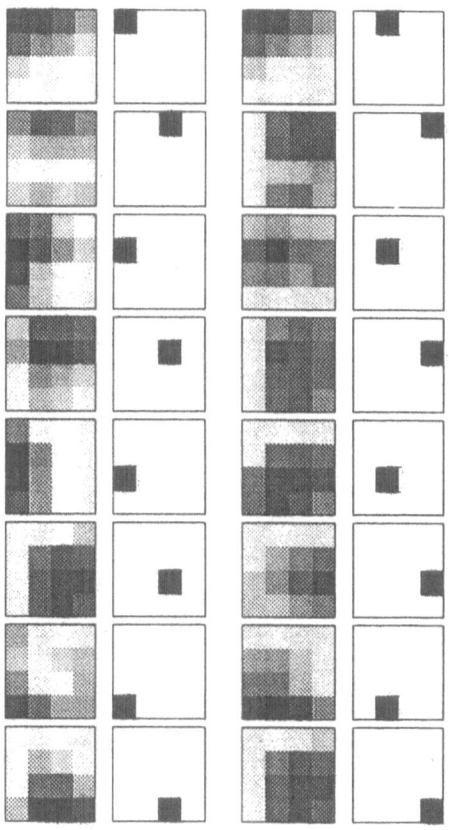

# An Apple Sorting System using Neural Network-Based on Image Processing

## Kazuho Kodaira

Systems Development Department
Sumitomo Heavy Industries Ltd.
2-4-15 Yato-Cho, Tanashi-City, Tokyo 188, JAPAN

## Hiroshi Nakata

## Matsuzo Takamura

## Abstract

An apple sorting system using neural networks was developed. The system determines three apple characteristics to inspect apple quality. They are surface colour grade, colour variation type and surface defects. They are judged by colour apple images taken with colour CCD cameras. The probabilistic neural network is used to determine its grade and the backpropagation networks are used for colour variation type and surface defects inspection. Throughput of 6 images per second are achieved.

# 1.Introduction

Quality of apples are judged by their appearances. Three apple characteristics are used for sorting:

Colour - Four colour grades
Colour Variation - Striped or Uniform type
Surface defects - Bruise, Hiyake, Sabi, Uneven Color

As these criteria are based on human sense, it's difficult to realize automated sorting system. The authors developed the prototype of automated apple sorting system using neural network-based on colour image processing as joint development with HNC Inc.

# 2.System configuration

Fig.1 shows the system configuration of this system. This apple sorting prototype system consists of three parts. First one is image capture system, second one is information processing subsystem and last one is apple transfer subsystem.

In the image capture subsystem, there are three colour CCD cameras, two strobe lights and an enclosure made of diffuser. Cameras are located to take top view, left and right side view images. Apples are put on the conveyer belt and moving on several hundreds mm/sec. Strobe lights are used because standstill images are needed to determine apple type and to detect surface defects.

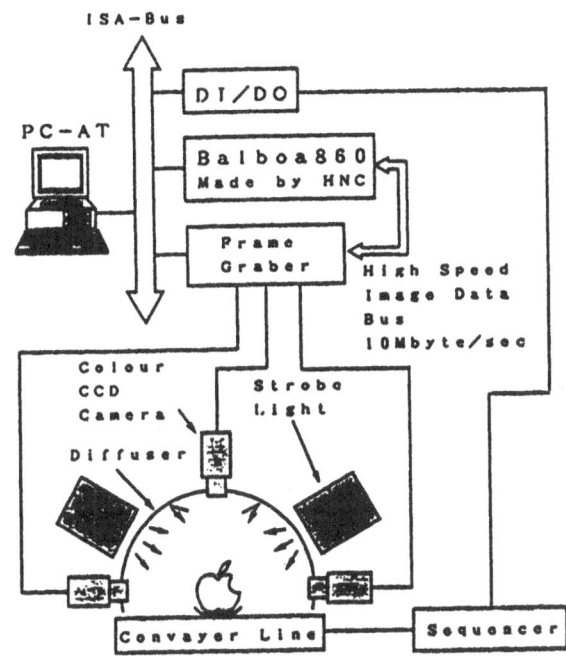

Fig.1 System Configuration

And also a diffuser enclosure is used to reduce glare and dark area on an apple image. Information processing subsystem is PC-AT computer and its add-on boards. Three boards are used; namely colour frame grabber boards, Balboa860 and DI/DO board. Frame grabber board has 512 by 480 pixel resolution. Three colour image(top and two side views) are subsampled on one image of this frame grabber. Balboa860 is neural network and image processing accelerator board developed by HNC Inc., using i860 processor. Almost all image processing and neural network algorithms are executed on this board. At high speed image data transfer is required to perform high throughput, direct image data transfer inter-

face between frame grabber board and Balboa860 are developed
for this purpose. This transfer rate is 10Mbyte/sec. DI/DO
board takes a communication role with sequencer of apple
transfer subsystem. Apple transfer subsystem consists of
conveyor belt and sequencer that controls a tray on the
conveyor belt to incline at the specified position for
sorting.

# 3.Sorting Algorithm

Fig.2 shows the block diagram of sorting algorithms. Sorting
algorithms are divided into four parts. These are image pre-
processing, colour variation type determination, colour
grade determination and surface defect identification. Each
algorithms are described briefly in the following sections.

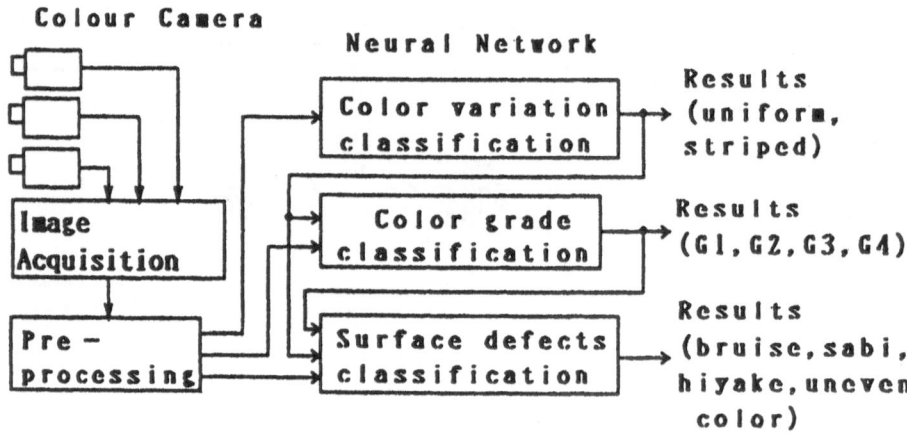

Fig.2 Block diagram of sorting algorithm

## 3.1 Image pre-processing

There are several processing steps that yield data which is
needed by more than one classifier. It is therefore efficient to do this processing only once. The pre-processing
takes are enumerated below.

(1)   Performing a RGB to Hue Saturation Intensity (HSI)
      conversion

(2)   Obtaining the Minimum Boundary Region (MBR) of the
      apple
(3)   Obtaining the redness and darkness values of the apple
      from the HSI values within the MBR

## 3.2 Colour variation type determination

The striped/uniform (type) determination system is run on
the first side view of each apple. This is because stripes
appear as vertically-oriented red information on a patch of
the image are processed to produce the feature vector.
Extracted feature vectors are as follows :

(1)   Redness
(2)   The standard deviation of the patch after stripe
      contrast enhancement
(3)   The number of significant extrema in the patch after
      stripe contrast enhancement

These feature vectors are fed into three layer Back Propagation Neural Network (BPN). This BPN is trained as type
determination network. Approximately one hundred apples are
needed for training.

## 3.3 Colour grade determination

The grade determination network yields the grade of the
apple using the top view of the apple. Feature vectors are
the redness, darkness and the type of the apple. Type is
used to select the network. Probabilistic Neural Network
(PNN) is used as grade classifier because it is needed that
the border of each grade can be moved easily. PNN has ability to move the border of the each class be altering the cost
function of each grade. About 20 apples for each grades are
needed to train the network.

## 3.4 Surface defect identification

Seven surface defect classes are defined : that is four
defects , glare, stem area and normal. Surface defect identification is performed with respect to each tile of input
image that contains apple imargery. There are two steps.
First step is to construct the defect map and second step is
post-processing.

Defect map is produced from the three layer BPN. There are separate BPN for top and side view, striped and uniform as well as grade 1-2 and grade 3-4 apples. The inputs to the BPN are values of colour elements (R, G, B) and relative distance of the tile from the centere of the apple. The output of the BPN is confidence value of the winning classification for that tile. This image is the defect map. These image is post-processed to yield the defects that exceed the minimum size specification.

User needs to cut out all kind of defect classes from several sample apples to train the BPN. Training time of this BPN is about 2 hours.

# 4.Discussion

Testing result of this prototype system is fairly well for grade and type classification. Throughput of 6 images per second (i.e., two apples per second) are achieved. But the improvement of image capture system is required because low quality of light on the edge of the apple causes defect miss-classification. And as higher throughput is needed for commercial system, we will try to reduce the algorithm execution time and to get higher detection speed.

# References

[1]     Rumelhart  D.E. and McCelelland J.L.,  "Parallel
        Distributed  Processing : Explorations in  the
        Microstructure of Cognition", I & II, MIT Press, 1986
[2]     Hecht-Neilsen R, "Neurocomputing", Addison-Wesley
        Publishing, 1990
[3]     Nakata H, Kodaira K, Takamura M, "An Apple Sorting
        System Using Neural Networks", SICE'91 Japan 611/612
        1991

# Linear Quadtrees for Neural Network Based Position Invariant Pattern Recognition.

Emmanouel C. Mertzanis, James Austin
Advanced Computer Architecture Group
Department of Computer Science
University of York
U.K.

## ABSTRACT

In the present paper a new method is proposed which involves teaching an Artificial Neural Network with a hierarchical data structure. This method results in a recognition process that significantly reduces the duration of a training process that is generally time consuming. The proposed method uses an invariant Octree data structure to describe the volume of a three dimensional object. The object is thus transformed in such a way that it is independent of rotation, translation and scaling and provides a significant storage capacity compression rate.

## 1. Introduction

The present paper aims to introduce a method capable of recognising an aircraft from any viewpoint in uncluttered noisy scenes. Current neural network based methods are limited by the time required to learn to recognise an object, as well as by the number of pictures required to be processed in order to achieve successful classification results. Typically, a number in the order of thousands of pictures per object can often be required in training multilayer neural networks. In this paper a new method that uses a single presentation of an image during training will be presented. The aim is to achieve the same successful position invariant classification results for three dimensional objects. The proposed method uses for this purpose a novel pre-processor, that is based on the properties of the Quadtree and Octree data structures combined with a neural network classifier. Both data structures have been extensively studied during the last decade [1-2], and have been shown to have two significant properties: firstly they can be made independent of the orientation, translation and scaling of the actual object they represent, and secondly they offer high compression ratios, normally in the range of 5:1 or 6:1; a considerable advantage when large numbers of digital pictures are to be stored within a storage medium.

This paper will be structured into five parts. A general description of the image processing procedures that were used during the input pre-processing steps, a detailed description of the Quadtree and normalised Quadtree data structures, the theory behind the Octree generation and the properties of that data structure, and finally the actual method used in order to train and test the artificial neural network with the specific results obtained in several cases.

## 2. Image Processing Procedures

Interest in digital image processing methods stems from two principal application

This project is partially funded by SERC.

areas: improvement of pictorial information for human interpretation, and processing of scene data for autonomous machine perception. Applications of digital image processing concepts did not become widespread until the middle 1960s, when third-generation digital computers began to offer the speed and storage capabilities required for practical implementation of image processing algorithms. Since then, this area has seen extensive growth, having been a subject of interdisciplinary study and research in many fields of science. In the current application the various image processing algorithms used, can be classified into two main categories. First, image enhancement algorithms that aimed to improve the original raster picture and second, image segmentation procedures that successfully clustered the enhanced image into meaningful regions for the later pattern classification stage. In image segmentation emphasis was placed primarily in edge detection operators and secondly in histogram thresholding techniques. A synoptic view of the overall image pre-processing procedure used in the present application can be viewed as a sequence of the following four steps: **a.** A 3 by 3 pixel Gaussian kernel was used to smooth the original raster image, and reduce any noise present in it. **b.** An edge detection segmentation procedure, based on the Maxgrad edge operator [3], was applied on the smoothed image. **c.** The resulting raster data were histogram normalised over the full 512 intensity range. **d.** The enhanced image from stage (c) was thresholded, thus eliminating the lowest 5% of edges (that normally correspond to noise). All remaining high grey scale intensities greater than zero were reduced to 1, resulting in the final binary edge map image. In the current application the raster images used, were images of six different aircraft types, created from precise detailed models and their corresponding pictures when the models were presented in front of photographic camera with a 512 by 512 pixel resolution [Figure 1]. A typical example of the above image processing techniques in the context of the current application can be seen in the following pages, where frames from an actual aircraft pre-processing sequence are shown.

## 3. Quadtrees

### 3.1. Data Structure Definition

Region representation is an important issue in digital image processing and

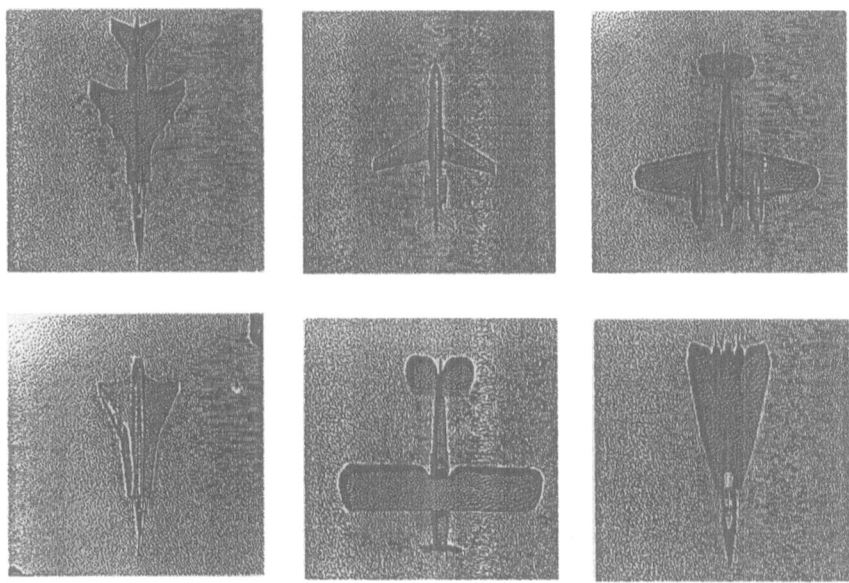

**Figure 1. The six aircraft models used in the experiments.**

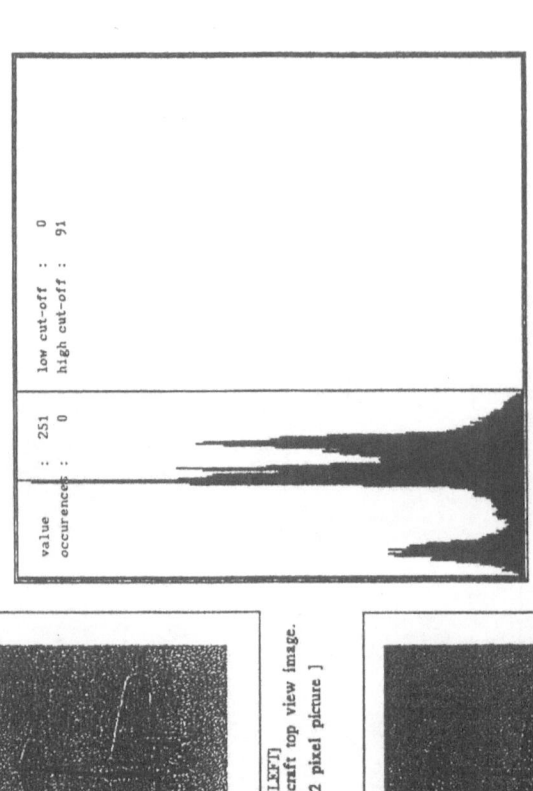

value        :  251        low cut-off   :   0
occurences   :    0        high cut-off  :  91

The histogram of the resulted Gaussian image.

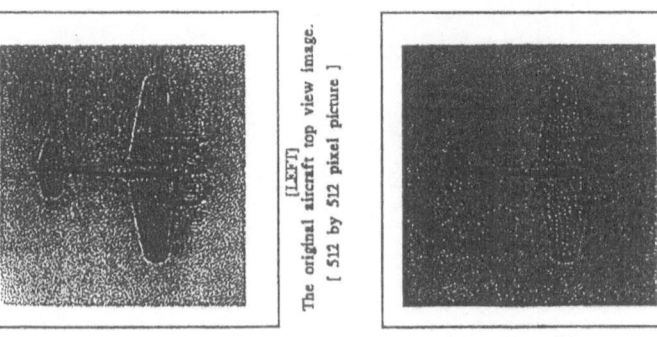

[LEFT]
The original aircraft top view image.
[ 512 by 512 pixel picture ]

[RIGHT]
The resulting raster image after
smoothing using a Gaussian
mask of size 3X3 pixels

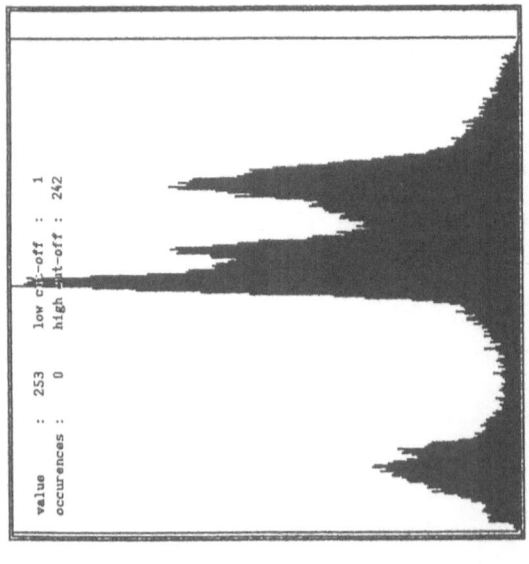

value        :  253        low cut-off   :   1
occurences   :    0        high cut-off  : 242

The histogram of the original top view aircraft image.

value : 255   low cut-off : 155
occurences : 1   high cut-off : 255

*Grey scale value* ☞ 0
*Occurrences* ☞ 35617

The histogram of the new edge map.

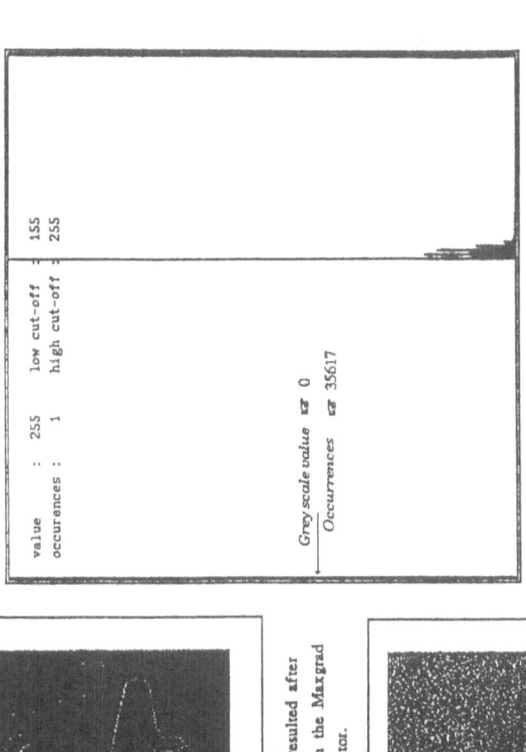

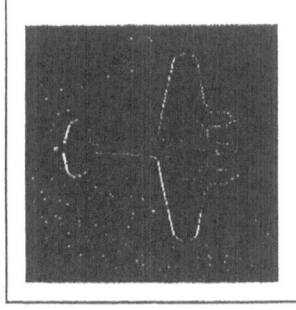

[LEFT]
The raster image resulted after edge detection with the Maxgrad edge operator.

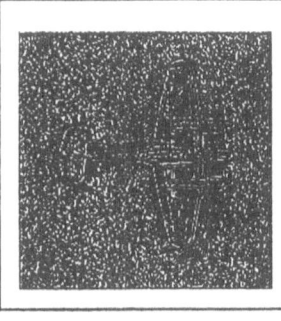

[RIGHT]
Raster image of the shifted edge map created by shifting the histogram towards the high intensities.
( threshold >= 155 )

value : 243   low cut-off : 0
occurences : 0   high cut-off : 148

*Grey scale value* ☞ 0
*Occurrences* ☞ 35617

The histogram of the edge detected aircraft image.

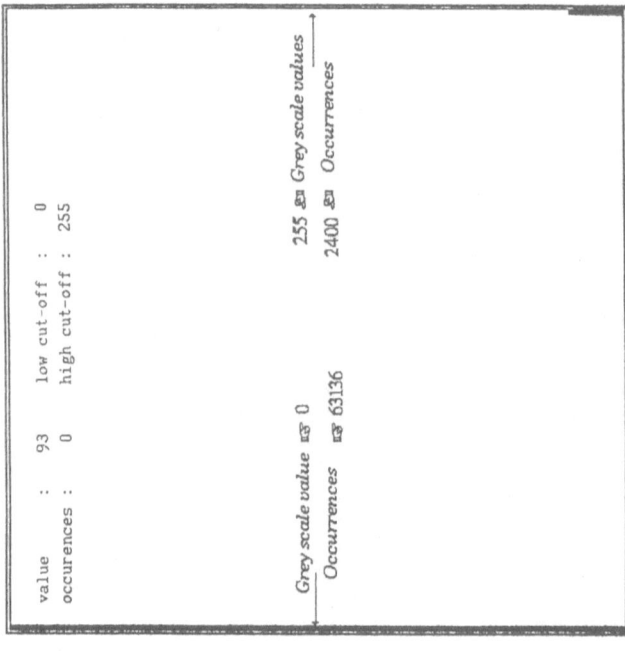

value      :    93     low cut-off  :    0
occurences :     0     high cut-off :  255

Grey scale value ☞ 0                          255 ☜ Grey scale values
Occurrences ☞ 63136           2400 ☜ Occurrences

**Above : The histogram of the final binary aircraft image.**
**Left Up: The picture after histogram thresholding ( thresh. <=165 )**
**Left Down: The picture after median filtering. (masks of size 2&4)**

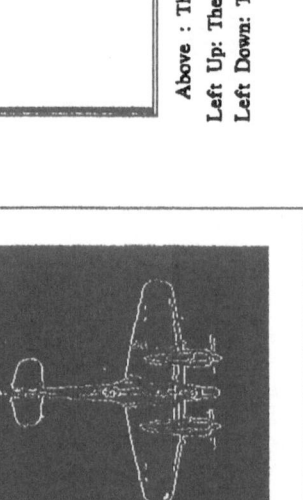

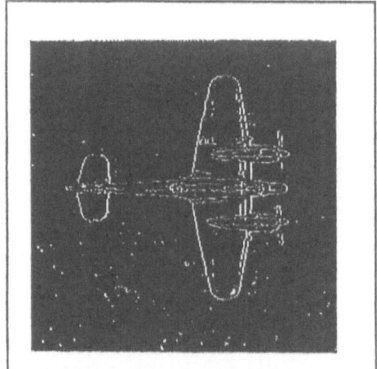

interactive computer graphics. Among the numerous representations currently in use, such as the run length technique, the chain codes and the binary trees, a recently developed idea by Klinger [4] has received considerable attention [Figure 2]. The Quadtree was originally devised as an alternative to the binary array image representation with the aim of saving space by grouping similar regions. However, more importantly, the hierarchical nature of the Quadtree also results in savings in execution time. In particular, algorithms for computing basic image processing operations using a Quadtree representation have execution times that are only dependent on the number of blocks in the image and not their size (e.g., for connected component labelling [5]). Given an image of size $2^N$ by $2^N$, a Quadtree representation can be constructed as follows: If every pixel in the image has the same value, the tree consists of a single (root) node labelled with that value. Otherwise, the image is divided into quadrants, and the root is given four children, representing these quadrants. The process is now repeated: If a quadrant consists entirely of one value, its corresponding node is labelled with that value; if not, it is split into subquadrants and its node is given four children representing them, and so on. The final result of this process is a tree of degree four (a "Quadtree") whose leaf nodes represent image blocks that have constant values. Clearly, the Quadtree representation described is likely to be quite uneconomical for grey scale images, in which large blocks of constant value will be rare. Therefore the current application examines only binary images as they might arise from segmenting a grey scale image.

Originally introduced by Klinger [4], the Quadtree has been undergoing a systematic evolution. Klinger and Rhodes [6] described Quadtrees by means of a systematic indexing of their elements with keys based on their node ancestors. Hunter and Steiglitz [7] applied a "rope net" consisting of pointers suitably linking a node to it neighbours. Samet [8][9][10], Dyer et al. [11] and Shneier [12] introduced a pointer based structure with a node linked only to its father and sons. Jones and Iyengar [13] presented a forest-like structure with a root and each tree addressed by a numeric key. In the general Quadtree implementation, the space required for pointers from a node to its sons is considerable. In addition, processing pointer-chains in external storage may be time consuming as a large number of page faults is often required. Consequently, there has been a considerable amount of interest in pointerless Quadtree representations. These can be grouped into two

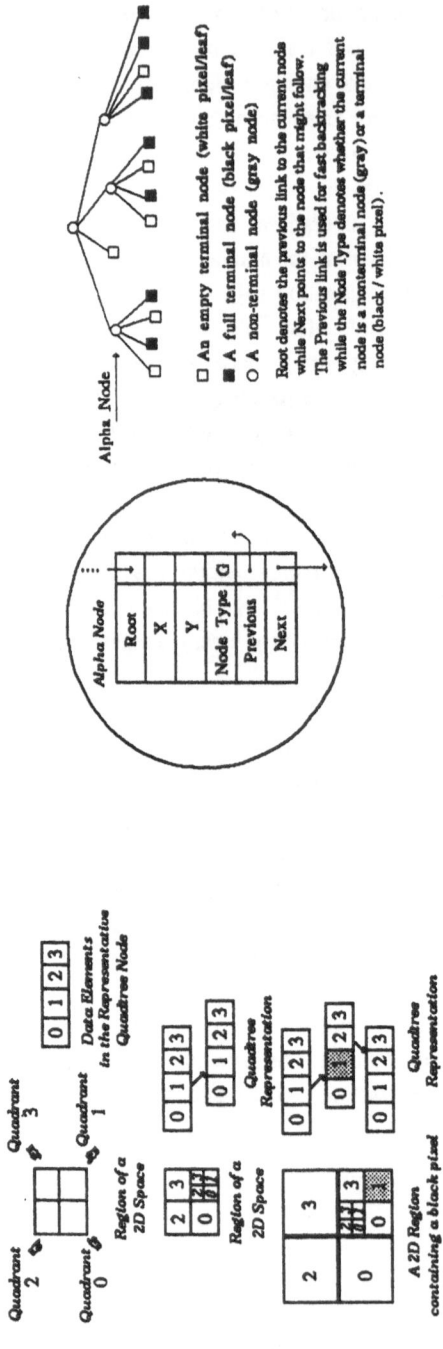

□ An empty terminal node (white pixel/leaf)
■ A full terminal node (black pixel/leaf)
○ A non-terminal node (gray node)

Root denotes the previous link to the current node while Next points to the node that might follow. The Previous link is used for fast backtracking while the Node Type denotes whether the current node is a nonterminal node (gray) or a terminal node (black / white pixel).

(Left) Figure 2. The Quadtree data structure.

(Right) Figure 3. The linked Quadtree data structure.

categories. The first treats the image as an ordered collection of leaf nodes. Each leaf is represented by a locational code corresponding to a sequence of directional codes to locate the leaf along a path from the root of the tree. Locational codes have been used by a number of researchers to represent Quadtree data structures [6] [14] [15] [16], as well as in other related contexts [17]. The second represents the image in the form of a preorder traversal of the nodes of its Quadtree [18].

Pointerless Quadtree representations such as those previously described are used in the current application and are generally known as linear Quadtrees. Their usefulness lies in their relative compactness and their appropriateness for maintenance of the data in external storage [Figure 3]. In linear Quadtree representation each pixel is encoded with a quaternary integer, the digits of which encode, from left to right, the successive quadrant subdivisions according to the following convention: 0 for the northwest; 1 for the northeast; 2 for the southwest and 3 for the southeast quadrant. For example consider the picture of Figure 4 and the corresponding linear Quadtree. Pixel Q=(0123) identifies a pixel which belongs to the northwest quadrant in the first subdivision, to the northeast in the second, to the southwest in the third and finally to the southeast in the last and fourth partition. Clearly, if four pixels such as 1000, 1001, 1002, and 1003 belong to the same quadrant relative to the $n$-partition , all are collapsed into one block, represented by a unique code, the last digit of which is a "wild card X". This process is often called condensation [19]. Clearly in the example of the previous figure, condensation would be applied a maximum of three times until 1XXX is obtained. A linear Quadtree allows an increase in the speed of data input, and retrieval from, the Quadtree. Links do not have to be processed, and it is often possible to traverse the tree with equal facility. An efficient addressing mechanism to realise these advantages is presented by Woodmark [14].

Quadtrees are hierarchical data structures used mainly for compact representations of two dimensional images. The advantage of the Quadtree representation for images is that simple and well developed tree traversal algorithms allow quick execution of certain operations, such as superposition of two images, area and perimeter calculation, and moments computation. Klinger and Dyer [20] have shown that the Quadtree representation of images yields substantial data compression over a variety of source images. In their

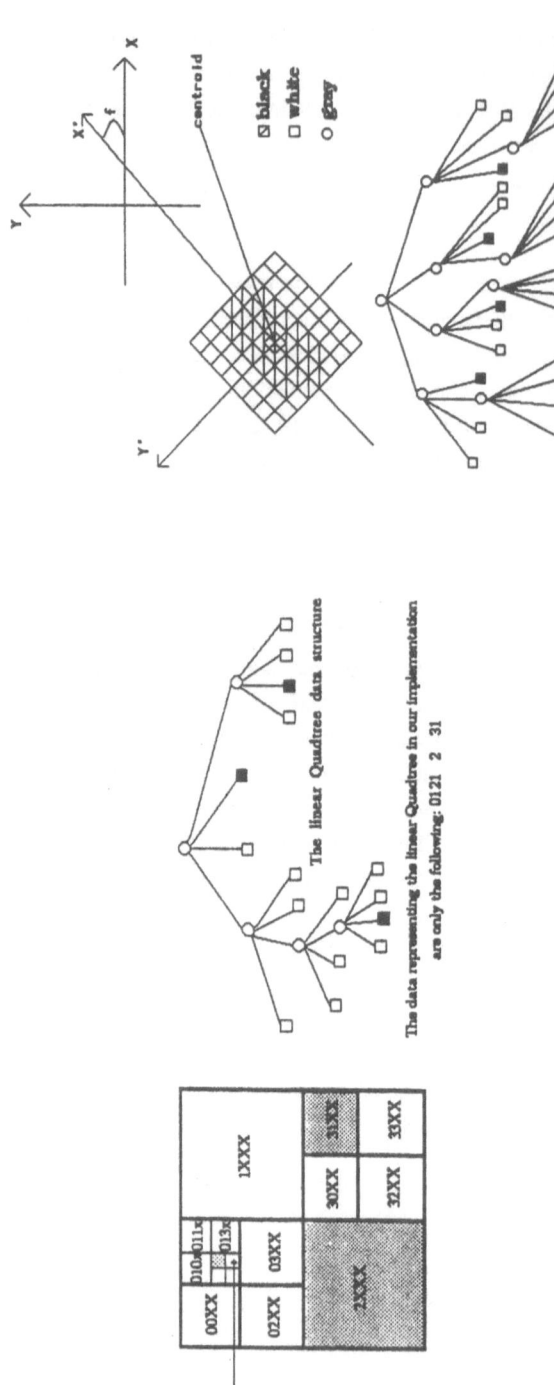

(Left) Figure 4. The linear Quadtree representation.

(Right) Figure 5. The Normalised Quadtree procedure and representation.

experiments, image compression ratios ranging between 3:1 and 33:1 were found, with 5:1 and 6:1 being the general compression factor. Tanimoto and Pavlidis [21] have additionally developed a recursive refining algorithm for edge detection utilising Quadtrees and have demonstrated its computational savings. However, the Quadtree representation has certain disadvantages. The Quadtree representation of an object in an image is heavily affected by its location, orientation and relative size. A small change in these parameters will generate different Quadtree data structures. To eliminate the effect due to the translation of objects in an image, Li et al. [22] defined a normal form of Quadtrees. They assert that this Quadtree representation is unique for any image under a class of translation. However, the problem arising from rotations and size change was not investigated. Recently, Chien and Aggarwal [23] proposed a representation scheme, the normalised Quadtree representation, that is invariant to object translation, rotation and size change.

### 3.2 The Normalised Quadtree Data Structure.

Instead of generating a Quadtree for the entire image, Chien and Aggarwal have created a normalised hierarchical data structure for each object in the image, the normalised Quadtree. The particular implementation of the normalised Quadtree data structure occurs as follows: The object is normalised to an object-centered coordinate system, with the centroid as the origin and its principal axes as coordinate axes, the object is then scaled to a standard size. In this way, the normalised Quadtree of an object is dependent only on the shape of the object, but not affected by its location, orientation, or relative size. With those invariant properties, the Quadtree of an object can be used as a shape descriptor to aid in the identification of objects in images if complete views of the objects are present and is an information preserving shape descriptor [24] [Figure 5].

## 4. The Octree Structure

### 4.1. Introduction

The availability of an efficient and complete representation of three dimensional objects is crucial to many computer vision and interactive computer graphics applications. All the various methods for three dimensional object representation may be categorised loosely into three main classes: volumetric descriptions, wire-frame descriptions, and surface descriptions. Nevertheless, excessive computation time and space requirements in data manipulation limit the applicability of many of the proposed techniques. An efficient three dimensional indexing scheme that greatly reduces the magnitude of these problems is the Octree representation scheme [25]. Octrees are developed as a means of making space array operations more economical in terms of memory space. They can be considered as the three dimensional extension of Quadtree methods. Using Octrees, objects can be represented compactly [26] and image processing tasks as computing volume or the moments of inertia matrix, can be efficiently implemented. Furthermore this new three dimensional modelling scheme and its associated linear growth algorithms allow objects of arbitrary complexity to be encoded, manipulated, analysed and displayed interactively in real time or close to real time in parallel, low cost hardware. The goals of Octree encoding scheme are as follows:

(1) To develop an ability to represent any three dimensional or N dimensional object to any specified resolution in a common encoding format.

(2) To operate on any object or set of objects with the Boolean operations (union, intersection, etc.) and geometric operations (translation, scaling, and rotation).

(3) To implement a computationally efficient (linear) solution to the N dimensional interference problem.

(4) To develop the ability to display in linear time any number of objects from any viewpoint with shading, shadowing, orthographic or perspective view and smooth edges (anti-aliasing).

(5) To develop a scheme that can be implemented across large number of inexpensive high - bandwidth processors that do not require floating point operations, integer multiplication or integer division.

### 4.2. The Octree Data Structure.

To create hierarchical indices for Octree representation, a cubic "object space" large enough to accommodate the object is considered. This is more often referred to as the "universal cube". The Octree is a regular cellular decomposition of the object space (universe). The universe is subdivided into eight cells of equal size. If any one of the resulting cells is homogeneous, meaning that it lies entirely inside or outside the object, the subdivision stops. If however, the cell is heterogeneous, that is, intersected by one or more of the object's boundary surfaces, the cell is sub-divided further into eight subcells. The sub-division process stops when all the leaf cells are homogeneous to some degree of precision. An Octree can be considered as a $2^n$ x $2^n$ x $2^n$ array of unit cubes. Each unit cube has associated with it a label, either full or empty. An object of diameter $2^n$ , where n>0, may be divided into eight cube-shaped arrays of diameter $2^{n-1}$. Any cubic region obtained in this manner will be referred to as an octant. Figure 6 shows the numbering of the octants. For an Octree representing an object array of diameter $2^n$, root is at level n, a node that corresponds to a single unit cube is at level 0, and a node is at level $i$ if it corresponds to an octant that contains $2^i$ x $2^i$ x $2^i$ unit cubes. For the object of Figure 7 the corresponding simple Octree structure is shown. An Octree representation usually reduces memory requirements over an object array. At the same time, it is more complex to access the data in Octree form. The geometric information contained in an Octree is implicit, and can be retrieved by use of procedures. Roughly, the location of an octant is derived by traversing the tree, and the size of the octant is determined by the level of the tree at which its corresponding node resides.

The Octree data structure for the representation of three dimensional objects has been under intense investigation for several years [26] [27]. An Octree can be generated from different forms of sensor data, for instance, intensity images from multiple views [23] or a series of sliced images acquired in computerised tomography [24]. Initially, the Octree representation was used to describe the volume occupied by a three dimensional object. Calbom et al. [28], proposed the polytree structure in which a leaf cell can be one of five types: full, empty, vertex, edge or surface. More recently, Chien and Aggarwal [23] proposed and implemented an algorithm to generate a Volume/Surface Octree from multiple views (silhouettes) of an object. The Volume/Surface Octree is an Octree

76

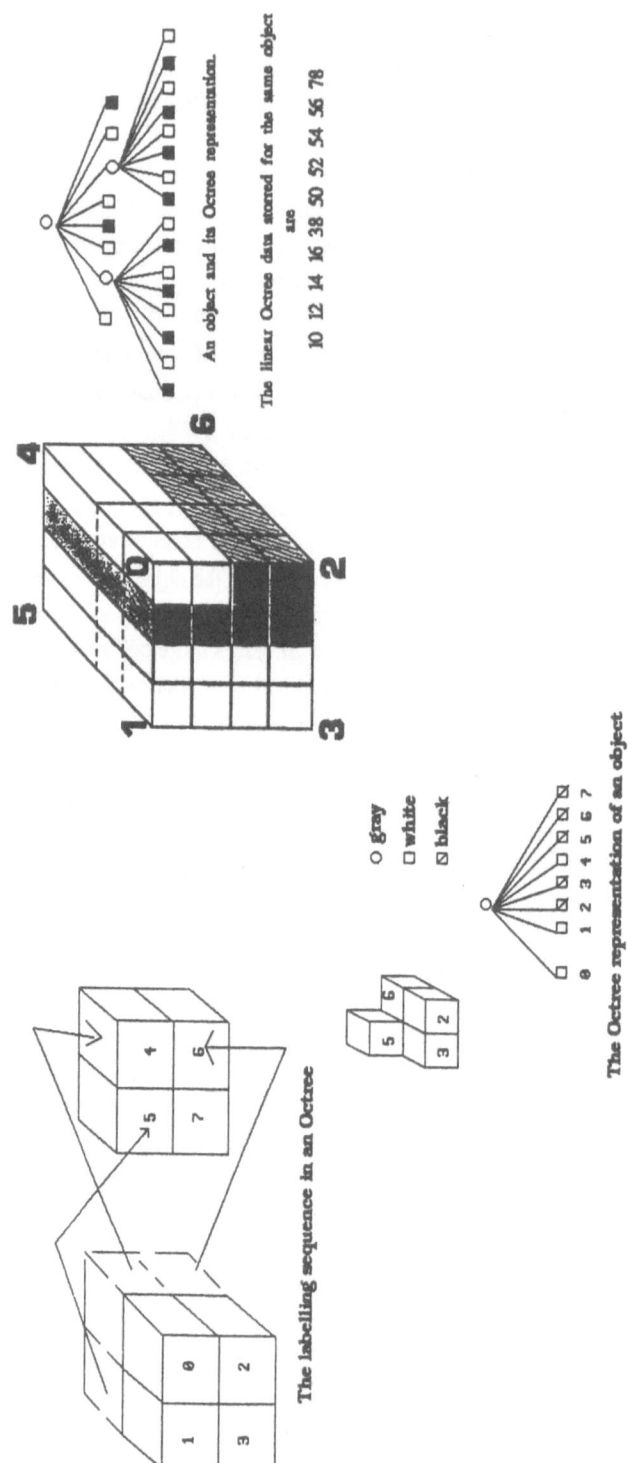

An object and its Octree representation.

The linear Octree data stored for the same object are

10 12 14 16 38 50 52 54 56 78

○ gray
□ white
☒ black

0 1 2 3 4 5 6 7

The Octree representation of an object

The labelling sequence in an Octree

(Top Left) Figure 6. Numbering the Octree's octans.

(Middle Bottom) Figure 7. An object and its corresponding Octree representation.

(Top Right) Figure 8. The linear Octree representation.

structure with information about the surface orientation encoded into its nodes. It is a compact and informative description of three dimensional objects, and is particularly useful in many computer graphics and object recognition applications. Moreover, due to its intrinsic tree structure, most algorithms performed on Octrees can be implemented on parallel computers.

The Octree's requirement of eight pointers from each internal node to its children makes its space requirements quite substantial. To overcome this disadvantage the concept of pointerless representation for Quadtrees, as well as for Octrees, has been developed [15] [29]. Here each leaf is represented by a locational key that is a numeric index obtained from the coordinates of an octant. Linear Octrees are useful because of their compactness and appropriateness in maintenance of data. In certain applications pointerless structures lead to a more space-efficient algorithm than their pointer-based counterpart. The present chapter aims at generating linear Octrees for three dimensional objects obtained by the volume intersection method. In similar cases where algorithms have been developed to implement the previously mentioned technique, it has been observed that pointerless representations were more efficient in space and time than an algorithm corresponding to pointer-based conventional Octrees [Figure 8].

# 5. The Volume Intersection Technique

## 5.1. Introduction

The problem of generating Octrees from a number of images was first addressed by Conolly [30]. He proposed a technique to generate the Octree of an object from an arbitrary view. The structure of the Octree could be cumulatively updated by subsequent views. In his approach the data for each view were first converted to a Quadtree. Since the coordinate axes associated with the Quadtree were not necessarily aligned with those of the Octree, the Quadtree needed to be transformed into the Octree coordinate system. However, the author failed to point out the complexity involved in the transformation (from the Quadtree to the Octree coordinate system in the case where the two coordinate systems are not aligned with each other), and did not provide enough justification for the conversion of input data to a Quadtree.

Generating Octrees from Quadtrees has been a very popular approach. It was first proposed by Yau and Shihari [24] to generate the Octree of an object from the Quadtrees of its cross sections. Later, Chien and Aggarwal [31] developed a technique to generate the Octree of an object from the Quadtrees of its three noncoplanar views, in which the "generalised" coordinate system of the Octree is specified by the three views. More recently, Veestra and Ahuja [32] have developed a technique to construct the Octree of an object from its "face" views", "edge" views, and "corner" views and Chien and Aggarwal introduced the idea of volume and surface Octrees from silhouettes of multiple views [33]. The later method can be classified into three main stages:

**a.** Obtain three images of the object using no coplanar views. Normalise each image using the method described in previous parts of this chapter, and create the respective normalised Quadtrees.

**b.** Use the normalised Quadtrees to generate three pseudo-normalised Octrees.

**c.** Use the Volume Intersection technique [34] (i.e., merge the three pseudo-volumes) and create the final normalised Octree structure.

Based on the above general algorithmic presentation, a more specific methodology has been formulated for the precise implementation of the above steps regarding each individual view. In this chapter the top, side and front views of the object were only used for convenience. In order to create the normalised Octree structure, the processing method followed for each particular view is described below in details. Figure 9(a) shows the original three views of a hypothetical 8 by 8 pixel aircraft model.

## 5.2. The front view image

The processing starts with a 2-D image that represents the front view of the object. No preliminary processing is required prior to the application of the Quadtree encoding

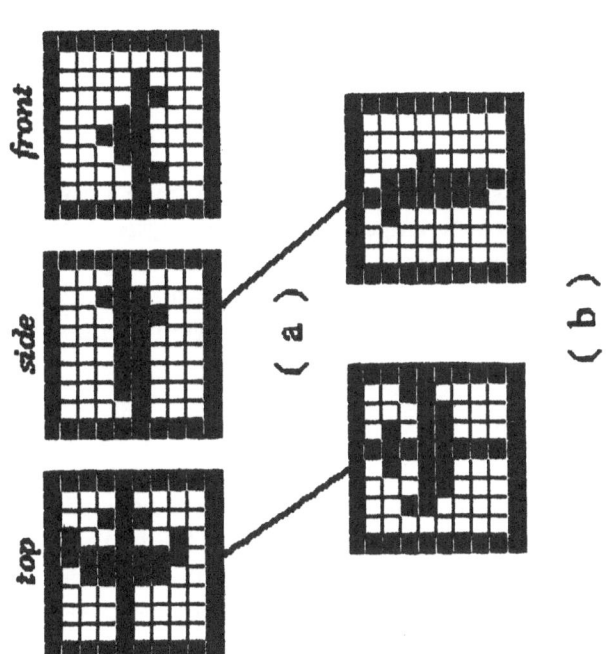

**Figure 9. The three original aircraft views and their initial pre-processing.**

algorithm [Figure 9(a)]. Assuming that the image is normalised, the method proceeds to create the normalised Quadtree representation of the image. The Quadtree structure is created by starting at the lower left quadrant and continues to the lower right, upper left and upper right, respectively. In order to create the Octree structure that corresponds to the front view volume of the object, we follow the next two rules: a) start from the root of the Quadtree and reverse all subsequent Quadtree nodes and leaves (e.g., consecutive node leaves a,b,c,d of the Quadtree will become d,c,b,a on the Octree) b) begin from the left most sub-Quadtree and moving to the right, for every Quadtree node copy all the four node's children so as to create the eight necessary Octree branches (e.g., d,c,b,a leaves of a Quadtree node will then becomes d,c,b,a,d,c,b,a in the final Octree node). Figure 11(a) shows the three dimensional volume created by the front view image, while Figure 12 displays the algorithmic conversion from the Quadtree data structure to the respective three dimensional Octree data structure.

### 5.3. The top view image

A preliminary processing step is required in order to create a new 2-D image by rotating the original 2-D top view image as described in Figure 9(b). The next step involves generating the Quadtree structure of the new top view image. In order to create the Octree structure each two consecutive nodes on the Quadtree are taken, and are then copied into the Octree as described in Figure 13. Finally figure 10(a) shows the volume of the top view image.

### 5.4 The side view image

The volume that corresponds to the side view image is displayed in Figure 10(b) while figure 14 demonstrates the algorithm for converting a Quadtree node into an Octree node in the side view picture case. It is important to notice that each node in the Octree is produced when each of the main four quadrants of the corresponding Quadtree is copied on the double number octant and also on the immediate successive octant (e.g., quadrant 2 is copied on octant four ($4=2x2$) and also on octant five [$5=(2x2)+1$]).

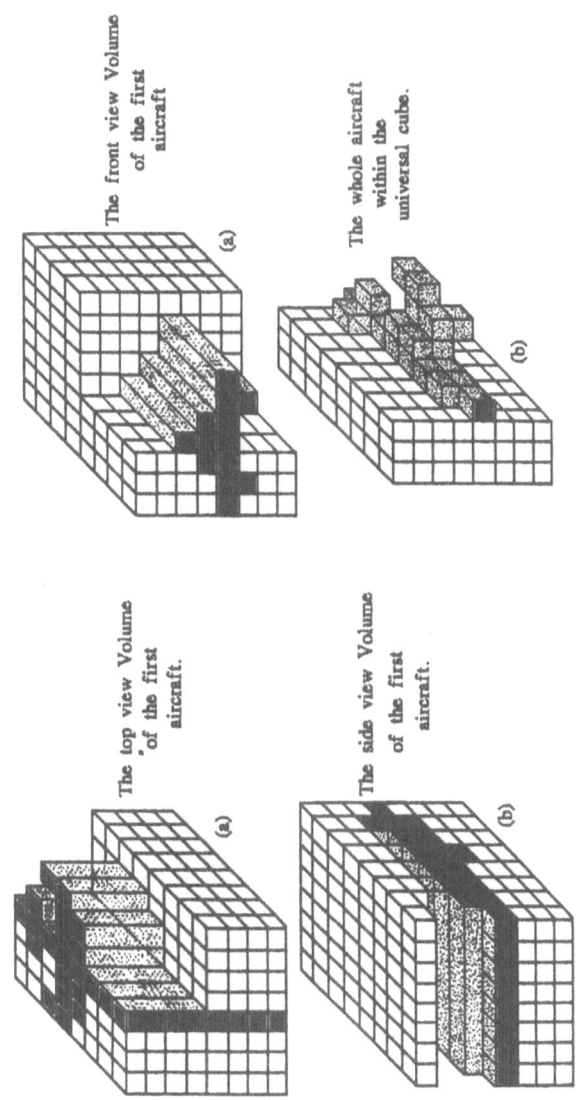

The front view Volume of the first aircraft

(a)

The whole aircraft within the universal cube.

(b)

The top view Volume of the first aircraft.

(a)

The side view Volume of the first aircraft.

(b)

(Left) Figure 10. The 3D volumes of the top and side view images.

(Right) Figure 11. Front view volume and final volume of the 3D aircraft.

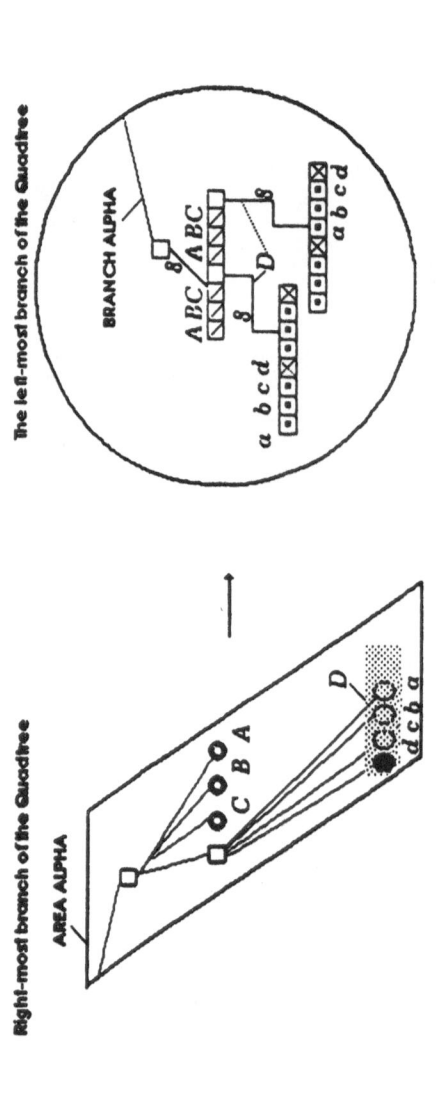

82

Right-most branch of the Quadtree

The left-most branch of the Quadtree

AREA ALPHA

BRANCH ALPHA

ABC ABC

ABC ABC

a b c d

a b c d

8

8

D

C B A

D

d c b a

(8) denotes that eight lines actually start from that point

**Creating the Pseudo-Octree of the Front View Image from its Quadtree Data Structure**

◆ a 2 by 2 white pixel area of the Quadtree
☐ an intermediate linking node of the Quadtree
○ a white Quadtree pixel
● a black Quadtree pixel

⊠ a 2 by 2 white octant area of the Octree
☐ an intermediate linking node of the Octree
⊠ a black Octree octant
⊡ a white Octree octant

**Figure 12. Converting a Quadtree node to an Octree node (front view).**

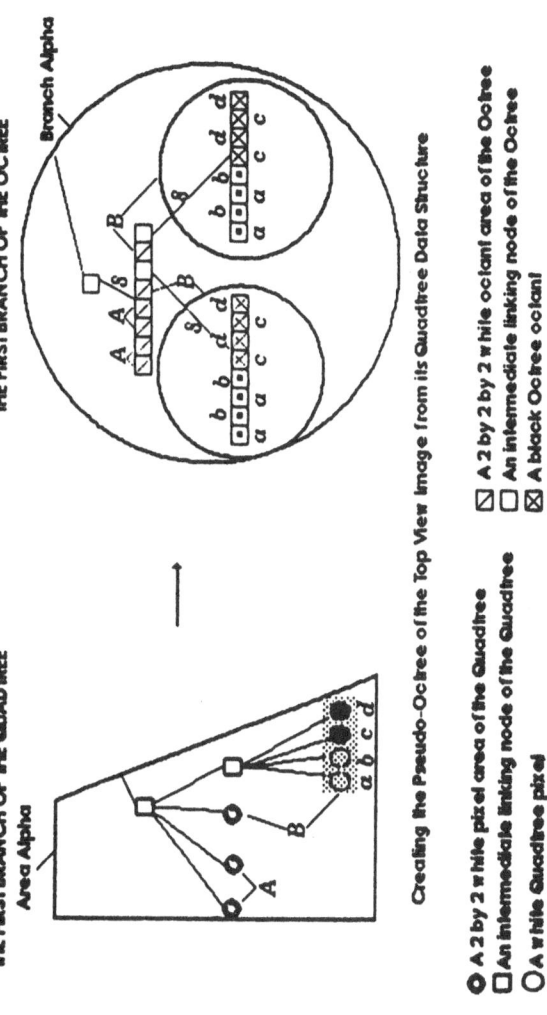

Figure 13. Converting a Quadtree node to an Octree node (top view).

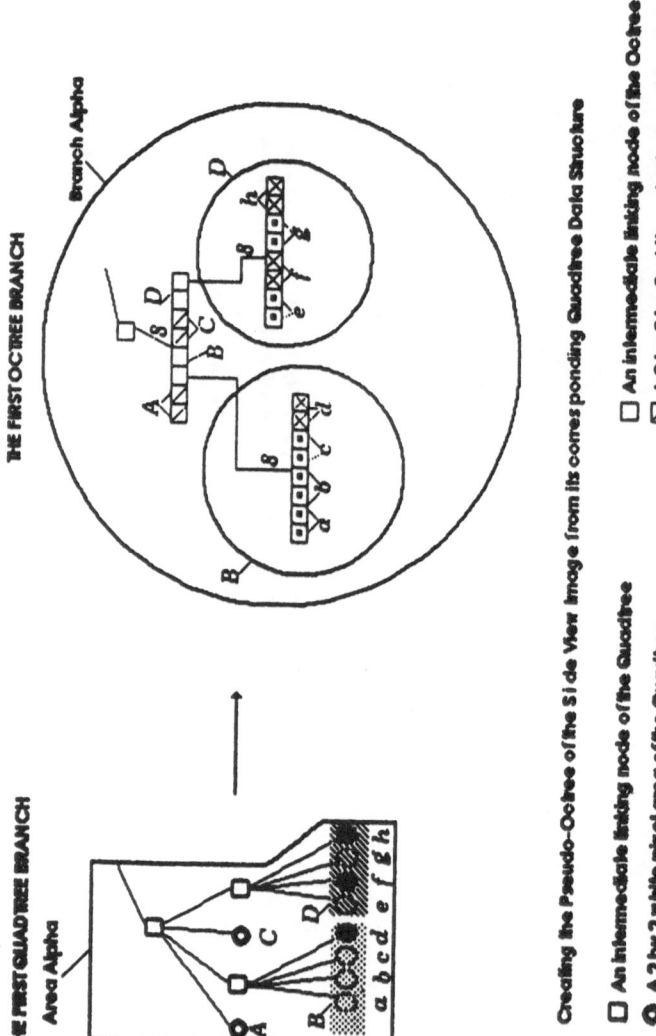

Figure 14. Converting a Quadtree node into an Octree node (side view).

### 5.5 The final Octree

In order to generate the final Octree structure that will describe the volume of the 3-D object of Figure 11(b), the technique known as Volume Intersection is applied [Figure 15]. Each one of the three normalised pseudo-Octrees is merged with another and the final node is full if all the nodes are full otherwise it is empty. Merging also involves the GRAY nodes. In the end, all nodes to be merged have to be of the same tree level and be either of full or empty type, so nodes may have to be expanded with the addition of more branches and further nodes in the Octree in order to obtain the same tree level for merging. In the example, the resulting normalised Octree is shown in figure 16 while the raster image used for training the ADAM neural network is shown in Figure 17. This synthetic image results from a simple in order traversal of the final Octree data structure.

# 6. Teaching the ADAM neural network.

The artificial neural network used here is a the Advanced Distributed Associative Memory (ADAM) [35]. ADAM artificial neural network shown in Figure 18, is a development of a correlation matrix memory, and can be viewed as a two layer network with a front end processor consisting of an N tuple processor. A full description of the network can be found in [35] along with an analysis of its storage ability. The following discussion describes the basic operation of the network. (The original description of ADAM does not describe it as a neural network. See [36] for a neural characterization of this network.) It possesses two important properties that together provide a high storage capacity in a small amount of storage space. Firstly, it uses two layers of weights (as opposed to one), and secondly, it explicitly code a 'class' pattern onto the middle layer of the network. Using this arrangement, the class pattern can be varied in size to accommodate different storage requirements. If only a single layer were used the capacity of the network would be enforced by the size of the input and the output pattern. To enable a simple and fast learning rule to be used the memory uses an N tuple pre-processing method. In the N-tuple method the pattern to be processed is placed in an array A [Figure 19]. This is then transformed into an array B by a number of decoders, $D_n$, which have the state transitions shown in the table of Figure 19. The contents of B are then

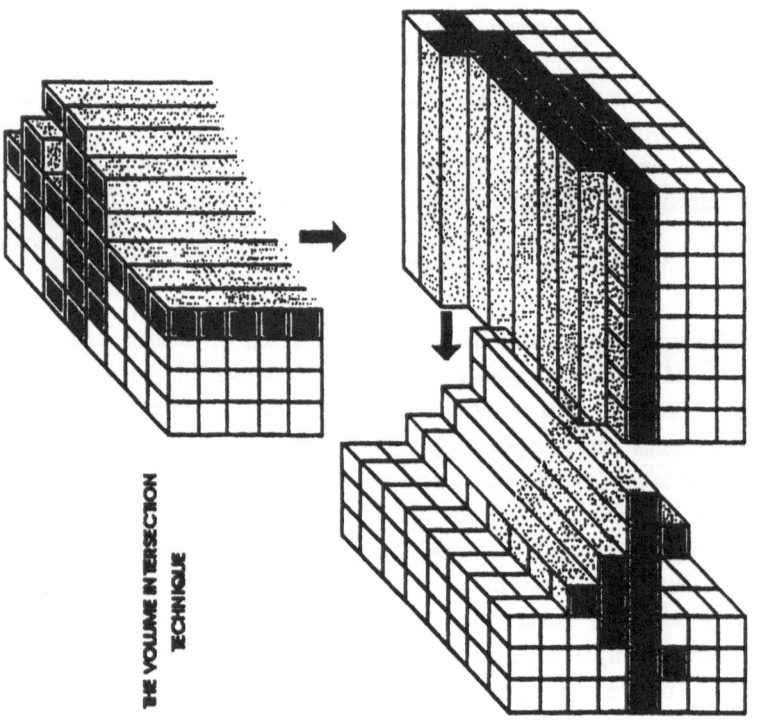

THE VOLUME INTERSECTION
TECHNIQUE

Figure 15. The Volume Intersection Technique.

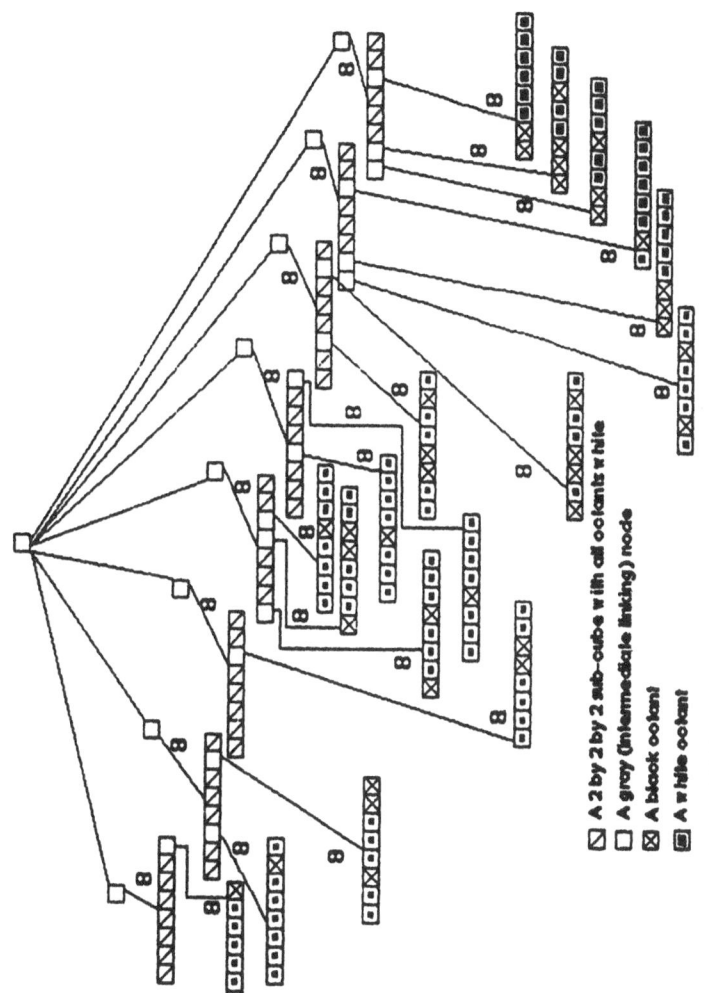

⊠ A 2 by 2 by 2 sub-cube with all colonts white
☐ A grey (intermediate linking) node
⊠ A block colonl
▣ A white colonl

THE OCTREE DATA STRUCTURE OF THE FIRST AIRCRAFT IMAGE

Figure 16. The final Octree data structure.

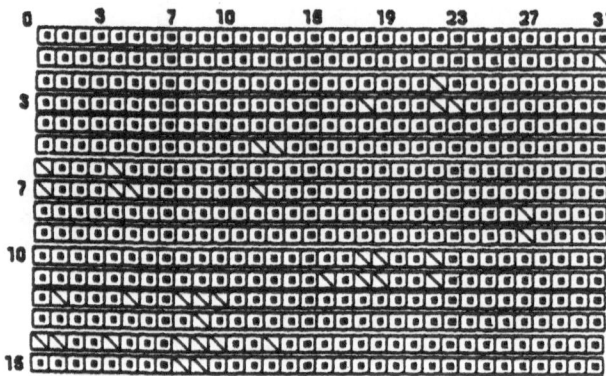

Figure 17. The synthetic image used for training ADAM neural network.

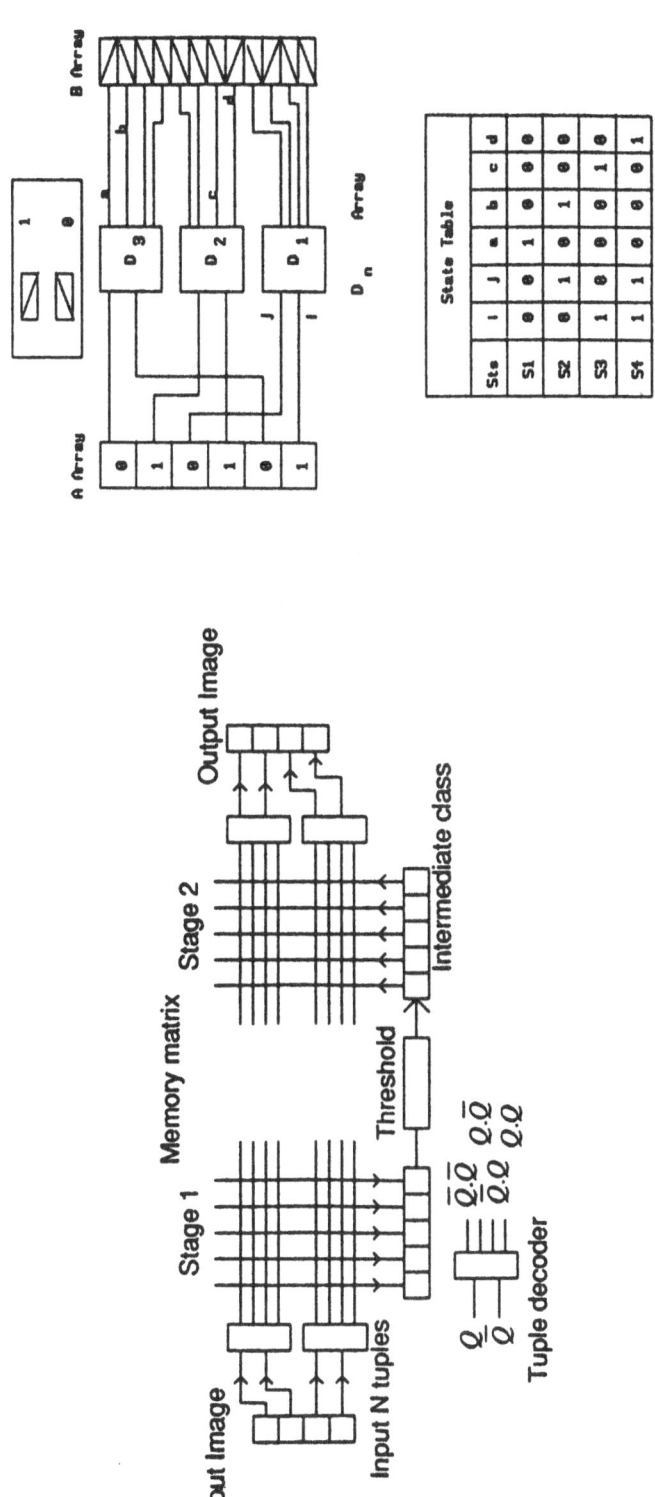

(Left) Figure 18. The ADAM neural network.

(Right) Figure 19: The N-tuple method.

fed into the first layer of the network. Two important points emerge from this process. Firstly, the pattern in A is converted to a pattern in B that will always contain $n$ bits at 1, where $n$ is the number of decoders. Secondly, the non-linear transform allows separation of non-linearly separable data. These properties allow teaching an association by a single presentation of an input output pair. Teaching is achieved by first correlating the input pattern with a "class" pattern using a correlation matrix and then the class pattern with the output pattern using a second correlation matrix [Figure 18]. For each new classification of an aircraft, a new class pattern is generated. This has $k$ bits set to one randomly distributed in the class array.

The weights in the memory are binary - if during training the input and the output units of a weight are both '1' the weight is set to '1', otherwise it remains at its current weight. Recognition in ADAM is similar to conventional networks. After N tuple pre-processing the image is presented to the first layer of the network. A weighted sum is computed for each class vector element. Once completed, the row class vector values are thresholded by selecting only the $k$ highest values to 1, all others to zero (where $k$ was the number of bits set in the taught class pattern). This process recovers the class, the identity of the image. This pattern is then presented to the next layer, and a weighted sum is performed to produce the output image. The row values in the output image are then thresholded, values of $k$ or greater are set to 1, all others to zero. The storage properties of this approach are quite efficient (see [35]). One of the main reasons for this is the ability to control accurately the number of links formed in the associative memory for every pair of patterns associated, in less than $n \times k$, where $k$ is the number of points set to one in a class pattern (constant), and $n$ is the number of points set to one in the input pattern after N-tuple pre-processing. The storage ability of the approach is increased by the use of the $k$ bit class pattern. By using this the neural network can efficiently use the correlation matrix and permit simple recovery of the class by $k$ point thresholding during tests.

In this application the second layer of ADAM is not exploited (although it may be used to store a prototype image of an aircraft). The class pattern is used to identify the current image. In order to teach ADAM, each generated Octree structure is modified so that each leaf of the octree lies on the upper level of the tree, (i.e., if the original image is of size 16 then all leaves are on level 4 of the tree, (16 = $2^4$)). Each voxel (volume element) is then used to formulate a 2-D binary image which characterises that particular octree structure. This image results from a rotation-, translation-, and scaling-independent hierarchical data structure, and it is the image used in order to teach the ADAM artificial neural network.

# 7. Results

The original aircraft images were 512 by 512 pixels, and were reduced to 64 by 64 pixels. Although clear structural details were lost, the reduction in size has resulted in substantial reduction of the noise levels present in the original picture and consequently higher probability for matching as well as structurally better images for training the neural network. The N-tuple size used in ADAM was four and the number of bits set in each class pattern was three. In addition, the class array size was 16 bits. The octree images used for training ADAM were all of size 512 by 512 pixels. The overall training time per image on a Sun 3/50 averaged about 50 seconds while retrieving associations took on average 45 seconds per image. The time required on average to read an octree image was 5 seconds. Only one synthetic octree image is required for each one of the six different aircraft types used in the experiments. The neural network has been trained originally with or without noise although in later stages and in order to investigate its generalisation properties ADAM has been trained both with noise as well as with pictures that have no added random noise. Two basic types of images have been used. Edge detected images with simple boundaries, and edge detected images with filled boundaries. For testing, a set of six images has always been used for each one of the six different aircraft types, resulting in an overall testing set of 36 images. The six images used for testing in each one of the six prototype aircraft included: two images (a,b) with no added random noise (having a significant number of deleted structural object pixels [15% on average]), and four images (c,d,e,f) of the original object volume with added random noise of 2%, 5%, 7% and 10%.

In the first set of experiments a set of only six octree images with no added random noise was used. When simple boundary pictures were used, ADAM showed a higher

92

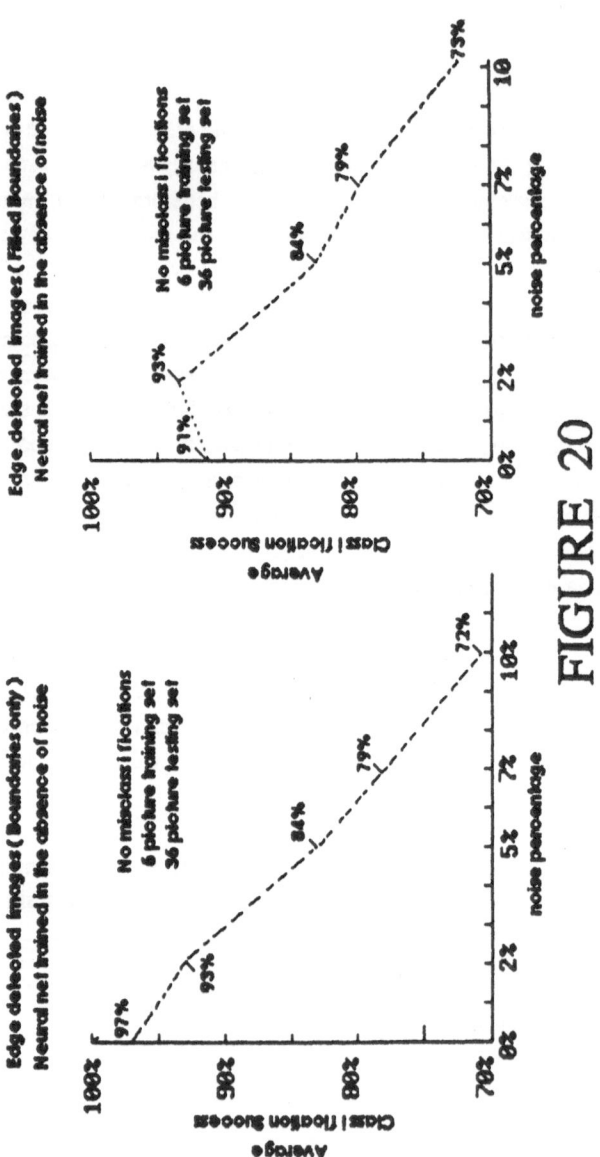

FIGURE 20

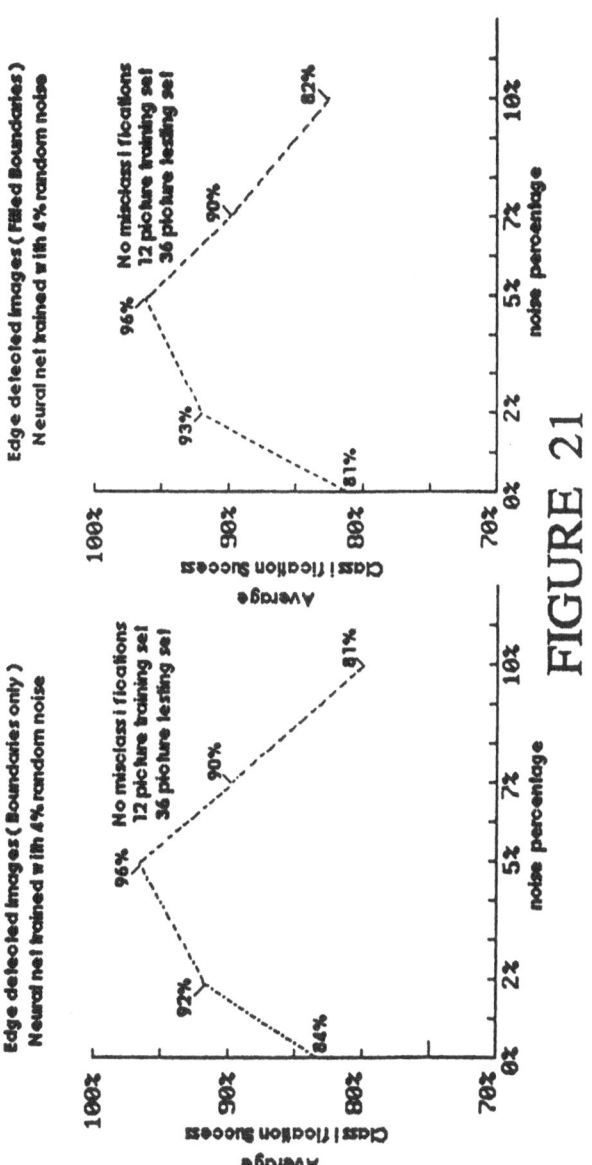

FIGURE 21

confidence towards the two pictures with no added random noise, while the confidence decreased as the level of added noise in pictures increased. The neural network showed a confidence ranging from 97% for the structurally distorted images (a,b), to 72% for the octree images with 10% added random noise (f). Aircraft classification has always been correct. When pictures with filled boundaries were used, ADAM showed a similar behaviour ranging from 92% for pictures (a,b), to 73% for the images with 10% added random noise (f) [Figure 20]. There was a small difference in that specific graph between the pictures (a,b) and the images with 2% added random noise (c). The latter are shown to have a higher confidence ratio than the former. The reason behind that anomaly is simply the fact that the percentage of object pixels distorted in pictures (a,b) was as much as 3.5%. Altering a large number of pixels on the filled boundaries of an object results in a form of localised noise addition, quite similar to a general random noise addition in an object's octree volume. In general, for the first set of experiments in the absence of any added random noise in the training set of six pictures, ADAM always showed greater preference towards the test pictures with the lower quantity of added random noise (localised noise). For the second set of experiments six octree pictures with 4% added random noise were used for training. When edge detected images with simple boundaries (skeletons) were used, ADAM showed higher confidence around the level of noise that was closer to the level of noise with which the neural network was trained. As a 4% added random noise was used for training, ADAM responded favourably at 5% added random noise in testing. The confidence curve shown in Figure 21 had a peak value at 5% added random noise and lower confidence ratio at (+/-) 5% from both sides of the peak value. ADAM was recorded to have a maximum confidence of 96% and an overall minimum value of 81% for the pictures with 10% added random noise (f). It is important to realise that there were no misclassifications. In the case of filled boundaries edge detected pictures, the neural network performed almost identically with its peak again found near the (b) images and the overall minimum value at 81% on the side with the (a,b) images. The reason for the later differentiation is the same as that described previously in the first set of experiments.

To conclude, for the second set of experiments when the ADAM was trained with 4% added random noise it performed most favourably with the images where the noise level was the same as the noise level it was trained with. In addition the overall confidence had improved significantly from the first set of experiments with the lowest confidence rate being above 80%. There were again no misclassifications. In order to test the generalisation properties of the current neural network a new set of experiments was designed for patterns with added random noise and patterns with no added random noise. In the first of two different sets of experiments pictures with 5% and 10% added random noise were used as well as the original octree image with no added random noise [Figure 22]. When edge detected images with simple boundaries were used, the network responded most favourably with those images that had the same noise levels as these used in the training image set (that is, 5% and 10%). The confidence was enhanced dramatically and in all cases was recorded to be above 95%. One misclassification occurred resulting in an overall average of 97% successful classifications. There were 18 images in the training set and 36 images in the testing set. Similar results were obtained when the neural network was trained with edge detected images with filled boundaries, except that a moderately lower confidence for the pictures with no added random noise was found that can be associated once again to their high ratio of localised noise. In the last set of experiments ADAM was trained with four pictures for each aircraft (a training set of 24 images). Three images were with added random noise (5%, 7% and 10%) and the last one was the original octree image with no added random noise [Figure 23]. In this last experiment the neural network showed an almost perfect confidence (100%) for the majority of images while for the pictures (a,b) the percentage in both edge detected simple boundaries pictures and filled boundaries pictures has been clearly above 95%. The only misclassification that has occurred was rather the neural network's inability to select the closest class pattern from a set of two space vectors that were very close to each other within the 3D vector space.

## 8. Conclusions

A method for position and scale invariant pattern recognition of three dimensional aircrafts has been presented. The method, which is based on the properties of Quadtree and Octree data structures, requires a single image of each aircraft for training and is capable of successfully classifying aircrafts if a single presentation of their corresponding raster volume images is used for training. If the original volume aircraft image is used together with other noisy aircraft pictures during training, the overall generalisation ability of the

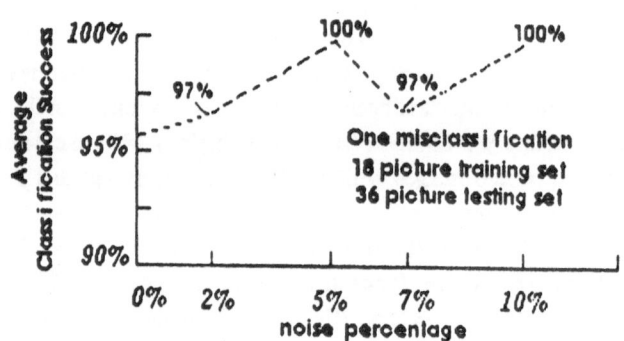

Edge detected images ( Boundaries only )
Neural net trained with 5% and 10% random
added noise

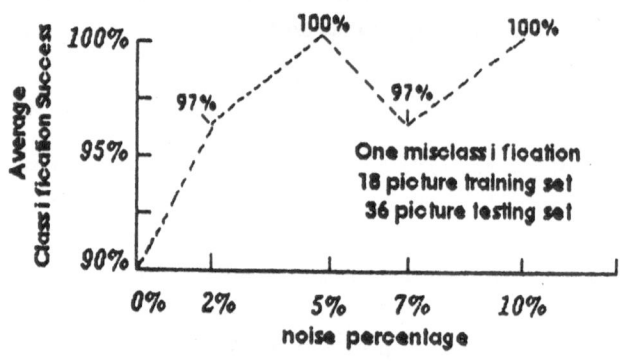

Edge detected images ( Filled Boundaries )
Neural net trained with 5% and 10% random
added noise

# FIGURE 22

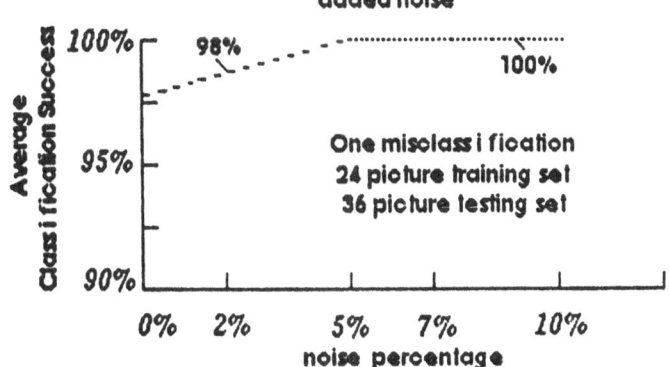

Edge detected images ( Boundaries only )
Neural net trained with 5%, 7% and 10% random added noise

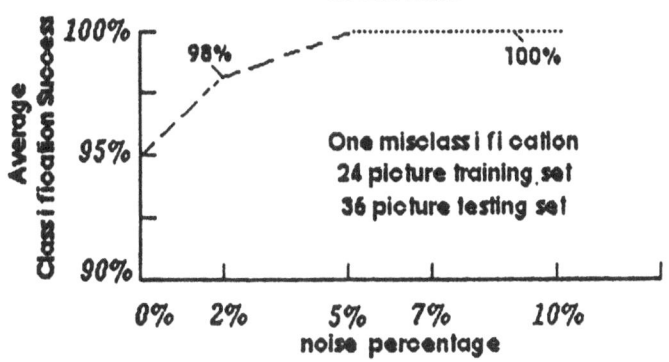

Edge detected images ( Filled Boundaries )
Neural net trained with 5%, 7% and 10% random added noise

# FIGURE 23

underlying neural network architecture is enhanced and the average confidence factor is shown to reach the maximum level.

## 9. Acknowledgements

The authors wish to acknowledge Mr. Davor Dukic who provided the DrawingBoard graphical tool that was used to design most of the enclosed figures.

# References

[1] A. Rosenfeld, " Quadtrees and Pyramids for pattern recognition and image processing", Proc. 5th Int. Conf. on Pattern Recognition, Miami Beach, 1-4 Dec 1980, 802-811.

[2] G.M. Hunter, "Linear transformation of pictures represented by Quadtree", Comput. Graphics Image Processing, vol 10, no 3, July 1979, 289-296.

[3] M. Imme, "A new method for edge detection", IEE 3rd Int. Conf. on Image processing and its applications, Number 307, University of Warwick, U.K., 18-20 July 1989, 656-659.

[4] A. Klinger, "Pattern and search statistics", Optimised Methods in Statistics, J.S. Ustagi ,ed., Academic Press, New York 1971.

[5] H. Samet, "Connected component labelling using quadtrees", J. ACM., vol 28, no 3, 1981, 487-501.

[6] A. Klinger, M. L. Rhodes, "Organization and access of image data by areas", IEEE PAMI-1, 1979, 50-60.

[7] G.M. Hunter, K. Steiglitz, "Operations on images using Quadtrees", IEEE PAMI-1, 1979, 145-153.

[8] H. Samet, "Region representation : Quadtrees from binary arrays", Comp. Graph. Image Proc., vol 13, 1980, 88-93.

[9] H. Samet, "An algorithm to convert rasters to quadtrees", IEEE PAMI-3, 1981, 93-95.

[10] H. Samet, "Neighbour finding in Quadtrees", Proc. PRIP-81, Dallas, Aug. 3-5 1981, 68-74.

[11] C. R. Dyer, A. Rosenfeld, H. Samet, "Region representation: boundary codes from

quadtrees", Comm. Ass. Comput. Mach. 23, 1980, 171-179.

[12] M. Shneier, "Path length distances for Quadtrees", Inf. Sci. 23, 1981, 45-67.

[13] L. Jones, S.S. Iyengar, "Representation of a region by a forest of quadtrees", IEEE PRIP-81 Conf., Dallas, 1981, 57-59.

[14] J. R. Woodwark, "The explicit quadtree as a structure for Computer Graphics", Comput. J., vol 25, no 2, 1982, 235-238.

[15] I. Gargantini, "An effective way to represent quadtrees", Comm. ACM, vol 25, no 12, 1983, 905-910.

[16] M.A. Oliver, N.E. Wiseman, "Operations on quadtree encoded images", Compt. J., vol 26, no 1, 1983, 83-91.

[17] G. M. Morton, "A computer oriented geodetic data base and a new technique in file sequencing", IBM Canada, 1966.

[18] M. Tamminen, "Encoding pixel trees", Comput. Vision Graph. and Image Proc., vol 28, no 1, 1984, 44-57.

[19] I. Gargantini, "An effective way to store quadtrees", Comm. Ass. Comput. Mach. 25, 1982, 905-910.

[20] A. Klinger, C.R. Dyer, "Experiments on picture representation using regular decomposition", Comp. graph. Image Proc., Vol 5, 1976, 68-105.

[21] S. Tanimoto, T. Pavlidis, "A hierarchical data structure for picture processing", Comput. Graphics and Image Processing, vol 4, 1975, 104-119.

[22] M. Li, W.I. Grosky, R. Jain, "Normalised Quadtrees with respect to translations", Proc. PRIP-81, Dallas, Texas, Aug. 3-5 1981, 60-62.

[23] C.H. Chien, J.K. Aggarwal, "Volume/Surface Octrees for the representation of three dimensional Objects", Comp. Graphics Image Proc., Vol 36, 1986, 100-113.

[24] C.H. Chien, J.K. Aggarwal, "Normalised Quadtree Representation", Comp. Vis. Graphics Image Proc., vol 26, 1984, 331-346.

[25] M.M. Yau, S.N. Srihari, "A hierarchical data structure for multidimensional digital images", Comm. ACM, vol 26, no7, 1983, 504-515.

[26] S.N. Srihari, "Representations of 3D digital images", Comput. Surveys 13, 1981, 394-424.

[27] C.L. Jackins, S.L. Tanimoto, "Quadtrees, Octrees, and K-trees: A generalised approach to recursive decomposition of euclidean space", IEEE PAMI-5, no 5, 1983, 533-

539.

[28] I. Calborn, I. Charkrarartn, Vanderschel, "A hierarchical data structure for representing the spatial decomposition of 3D objects", IEEE Comp. Graphics Appl. 5, no 4, 1985, 24-31.

[29] D. J. Abel, J. L. Smith, "A data structure and algorithm based on linear key for rectangular retrieval problems", Comput. Vision Graphics and Image Proc., vol 24, 1983, 1-13.

[30] C.I. Conolly, "Cumulative generation of Octree models from range data", Proc. Int. Conf. Robotics, Atlanta, March 13-15, 1984, 25-32.

[31] C. H. Chien, J. K. Aggarwal, "Identification of 3D objects from multiple silhouettes using Quadtrees/Octrees", Comput. Vision Graphics and Image Proc. 36, 1986, 256-273.

[32] J. Veestra, N. Ahuja, "Efficient Octree generation from silhouettes", Proc. IEEE Comp. Soc. Conf. on Comp. Vision and Pattern Recognition, Miami Beach, June 22-26, 1986, 537-542.

[33] C.H. Chien, J.K. Aggarwal, "Computation of Volume/Surface Octrees from Contours and Silhouettes of Multiple Views", Proceedings CVPR'86, IEEE Computer Society, Jun. 1986, 250-255.

[34] A.A.G. Requicha, "Representations for rigid solids: Theory, methods, and systems", Comp. Surveys, vol 12, no 4, 1980, 437-464.

[35] J. Austin, T.J. Stonham, "An associative memory for use in image recognition and occlusion analysis", Vision and Image Computing, Vol 5, no 4, Nov. 1987, 251-261.

[36] J. Austin, "ADAM: An Associative neural architecture for invariant pattern classification", IEE Conference on Artificial Neural Networks, IEE Conf. Publication 313, London 1989, 196-200.

# Application of Neural Networks in Process Control

N.A. Jalel    A.R. Mirzai*    J.R. Leigh    H. Nicholson**

Industrial Control Centre, The Polytechnic of Central London
London, U.K.

*Dept. of Computer Eng., Univ. of Science and Technology
Teheran, IRAN

**Emeritus Professor, Dept. of Control Eng., Univ. of Sheffield
Sheffield, U.K.

## Abstract

In process control, neural networks may be applied in imitating a skilled process operator, learning to represent a realistic cost function for optimal control, fault and malfunction diagnosis, and process modelling. The possibility of using artificial neural networks for fault and accident diagnosis in the Loss Of Fluid Test (LOFT) reactor, a small scale pressurised water reactor, is examined and explained in the paper.

## 1   Introduction

Artificial neural networks, which are referred to as neural networks, connections, adaptive networks, neurocomputers and parallel distribution processors, are massively parallel interconnected networks of simple elements intended to interact with the real world in the same way as biological nervous systems.

They have been used to solve a wide variety of science and engineering problems that involve extracting useful information from complex or uncertain data [1,2]. The back propagation algorithm has been used effectively to control a space lander simulation where the space craft should be landed before the fuel runs out and with as little terminal velocity as possible [3]. Another application of the back propagation neural net is for the dynamic modelling and control of chemical process systems where it has been applied to model the dynamic response of pH in a stirred tank reactor [4]. In the field of fault detection and diagnosis, the neural network approach has been used to diagnose faults in a three continuous-stirred tank reactor using a multi-layer feedforward perception model [5]. Neural networks have also been used as a connectionist expert system knowledge base for medical diagnosis [6,7]. In the field of nuclear reactors, neural networks have been used with the back propagation algorithm applied to identify four different abnormal behaviour patterns in the response of a simulated steam generator [8].

In this paper, a brief description of a nuclear power plant will be given, the concept of artificial neural networks will be explained, two different approaches for their use for fault diagnosis in the LOFT reactor [14,15] will be illustrated, and finally the two techniques will be examined and discussed through a hypothetical accident.

## 2   Nuclear Power Plant

The nuclear power plant is a very complicated system and there are numerous variations in the design and components of such systems. The plant contains an active core in which fission is sustained and in which most of the energy of fission is released as heat. The core contains the nuclear fuel, consisting of a fissile nuclide and usually a fertile material in addition and the moderator in most cases to slow down the neutrons which will be absorbed to achieve the fission. The core is surrounded by a neutron reflector to decrease the loss of neutrons from the core by scattering back many of those which have escaped. Usually the same material which is used as a moderator is used as a reflector and sometimes as a coolant. The coolant is pumped through the core to remove the heat generated in the reactor core due to the fission occuring and to take it to the boiler or heat exchanger in which steam is raised. The resulting steam or hot gas can then be used in a conventional turbine-generator system to produce electric power. This study has concentrated on the Loss of Fluid Test (LOFT) reactor which is a scale model of a large commercial Pressurized Water Reactor (PWR) producing 50 thermal megawatts.

## 3   Neural Network Model for Fault Diagnosis

Neural networks use the massively parallel-distributed processing potential of computational hardware and are aimed at developing information processing that is analogous to the biological nervous system [9,2,10].

The main unit of an artificial neural network is the processing element. Processing elements within a neural network are grouped together to form a structure called a layer, hence neural networks emerge from the interconnection of one or more layers of processing elements. A typical neural network usually consists of one or more intermediate or hidden layers of processing elements which are responsible for additional information processing with the input and output layer. The input layer receives input from the external world while the processing elements in the output layer have outputs that are taken to be the outputs of the network as a whole and these deliver the representation of the input after processing has occurred.

In a neural network model, learning deals with the ability of processing elements to modify and find the weights that will produce the desired behaviour where the learning algorithms usually work from the training examples or experience.

Back propagation and the threshold logic unit are the learning algorithms adopted in this work and are used for fault diagnosis in the LOFT reactor.

# 4 Back Propagation Technique for Fault Diagnosis

## 4.1 The Back Propagation Algorithm

The back propagation algorithm is the most common learning algorithm which has been tested with a number of different problems and has been found to perform well in most cases and to find good solutions. Back propagation is a learning algorithm designed to solve the problem of choosing weight values for a three layer artificial neural network with feedforward connections from the input layer to the hidden layer and then to the output layer. The back propagation algorithm performs the input to output mapping by minimizing a cost function using a gradient search technique, where the cost function, which is equal to the mean square difference between the desired and the actual net output, is minimized by making weight connection adjustments according to the error between the computed and desired output processing element values [11,12,13].

## 4.2 Network Design and Structure Using the Back Propagation Algorithm

A neural network technique will be used to assist the operator in diagnosing and identifying any abnormal behaviour in the LOFT reactor, by modelling the effects of the different accidents and faults that might occur. The back propagation algorithm is applied to train the network to identify and diagnose any fault by choosing the appropriate connection weights and internal thresholds for the processing elements.

The inputs for the networks are the symptoms for the different faults and accidents. However most of these symptoms are identified through monitoring the state variables of the LOFT reactor model. The state variables of the LOFT reactor model are analysed for any deviation from the normal operating limit which could happen due to increase or decrease in the values of the state variables due to abnormal behaviour in the reactor. For each state variable, two limits have been designed to identify the increase/decrease and further rate of increase/decrease from the normal operating limits. Another set of symptoms which involves information about the whole power plant is obtained and entered to the system by asking the user about their occurrence.

The input values for the network are discrete, taking on values of 1 and 0, where 1 refers to the occurrence of the symptoms which could be a deviation in the value of the state variable. This value could also be a sensor alarm received by the operator in the control room where the user of the system records the occurrence of the symptom. A zero value indicates that the system is in normal operating condition.

For the network with four processing elements in the input layer, as shown in figure 1, the input of the network is a set of four elements, one element for each node, and each element has a value equal to 1 or 0, thus there are a possibility of 16 sets of data which can be entered to the network ranging from (0 0 0 0) which indicates reactor normal operation to (1 1 1 1) which refers to the occurrence of all the symptoms of the accident.

The desired output of the network ranges between 0 and 1 depending on the state of the input set. When all the elements of the input set are equal to 0 then the desired output should be equal to 0 referring to normal operation and indicating that the percentage of faults occurring is 0 %. When all the elements of the input set are equal to 1, the desired output is given as equal to 1, indicating that the fault has occurred with a percentage possibility equal to 100 %. However, when one element of the set is equal to 1 and the rest are 0 the desired output is given as equal to 0.2 indicating a 20 % possibility of the fault occurring. Thus, in the case of two elements of the set equal to 1, the desired output is equal to 0.4 indicating a 40 % possibility of a fault occurring, while in the case of three elements are equal to 1 the desired output is given to be equal to 0.7 recording that the possibility of a fault occurring is 70 %. Table 1 illustrates all the possibilities of the input data to the network with the desired output value for each possible input set.

| The input sets | The desired output |
|---|---|
| (0 0 0 0) | 0 |
| (1 0 0 0) | 0.2 |
| ⋮ | ⋮ |
| (0 0 0 1) | 0.2 |
| (1 1 0 0) | 0.4 |
| ⋮ | ⋮ |
| (0 0 1 1) | 0.4 |
| (1 1 1 0) | 0.7 |
| ⋮ | ⋮ |
| (0 1 1 1) | 0.7 |
| (1 1 1 1) | 1.0 |

Table 1: The input data for four processing elements in the input layer

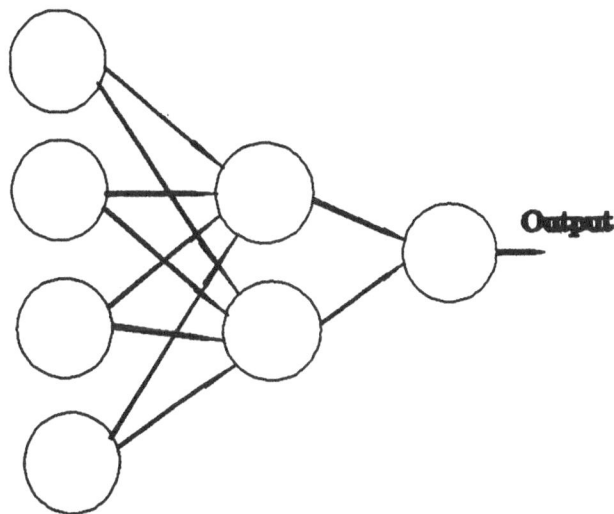

Figure 1: Network with four processing elements in the input layer

## 4.3 Network Performance Using the Back Propagation Algorithm

The computer program simulates the LOFT reactor model, monitors the state variables to identify any deviation from the normal operating limit, asks the user about the availability of some other information concerning the whole reactor plant and arranges and inputs the complete information to the networks. The output of the program is a list of the percentage possibilities of fault occurrence in the LOFT reactor. The program is designed to simulate the model, analyse the state variables, input the data to the networks and obtain the percentage list for all faults and accidents, at each second interval.

Partial loss of flow due to blockage in a fuel assembly is assumed to occur in the LOFT reactor to test and examine the ability of the program to diagnose the reactor accident. The hypothetical accident is assumed to be initiated by a drop in the coolant flow from 3.7 (lbm/hr) to 1.6 (lbm/hr) at a time greater than 2.0 sec.

As shown in figure 2, the indication for this accident involves four possible sets of symptoms which are

1. Decrease in coolant flow, change in flow distribution, increase in reactor coolant average temperature and change in temperature distribution.

2. Decrease in coolant flow, change in flow distribution, increase in reactor coolant average temperature and change in power distribution.

3. Decrease in coolant flow, change in flow distribution, increase in temperature difference across reactor core and change in temperature distribution.

4. Decrease in coolant flow, change in flow distribution, increase in temperature difference across reactor core and change in power distribution.

The network with four processing elements in the input layer is used to model this accident where the network is used four times, once for each set of symptoms, resulting in four possible output values in which the highest value will be considered as the network output.

Table 2 illustrates the percentage possibility of all faults and accidents in the reactor at times 0, 3, 4 and 5 sec. At time 0 sec, the maximum network output or fault percentage is 2.8 % revealing that the reactor is in a normal situation. At time 3 sec, the accident has occurred, the faults percentage for the partial loss of flow due to blockage in a fuel assembly has the highest accident percentage which is 19.9 %. At times 4 sec and 5 sec the partial loss of flow also has the highest percentage value where at time 5 sec the percentage value is 96.3 % meaning that the network output is near to the desired output 1, indicating that all the symptoms for the accident have occurred and the network input set is (1 1 1 1).

| Accident type | Percentage failure | | | |
| --- | --- | --- | --- | --- |
| | Time | | | |
| | 0 | 3 | 4 | 5 |
| Failure in power control system to respond to boron increase | 2.8 | 2.8 | 31.04 | 50.07 |
| Failure in power control system to respond to boron decrease | 2.8 | 2.8 | 2.8 | 2.8 |
| Failure in the rod drop | 2.7 | 2.7 | 19.9 | 19.9 |
| Failure due to misalignment of RCCA (Reactor Control Cluster Assembly) | 2.8 | 2.8 | 2.8 | 2.8 |
| Leakage in feedwater piping | 2.8 | 2.8 | 2.8 | 2.8 |
| Blockage in feedwater piping | 2.8 | 2.8 | 9.7 | 9.7 |
| Excessive feedwater heat removal | 1.7 | 1.7 | 1.7 | 29.9 |
| Blockage in fuel assembly | 2.7 | 19.9 | 70.3 | 96.3 |
| Loss of coolant | 1.7 | 1.7 | 1.7 | 1.7 |
| Failure in the pressuriser heater control | 1.7 | 1.7 | 1.7 | 1.7 |
| Failure in the pressuriser spray control | 1.7 | 1.7 | 1.7 | 1.7 |
| Failure in the pressuriser water level control | 2.7 | 2.7 | 2.7 | 2.7 |
| Steam generator tube rupture | 1.7 | 1.7 | 1.7 | 1.7 |
| Steam line break | 2.8 | 2.8 | 2.8 | 2.8 |

Table 2: The fault percentage for the different faults in the reactor at time 0 sec

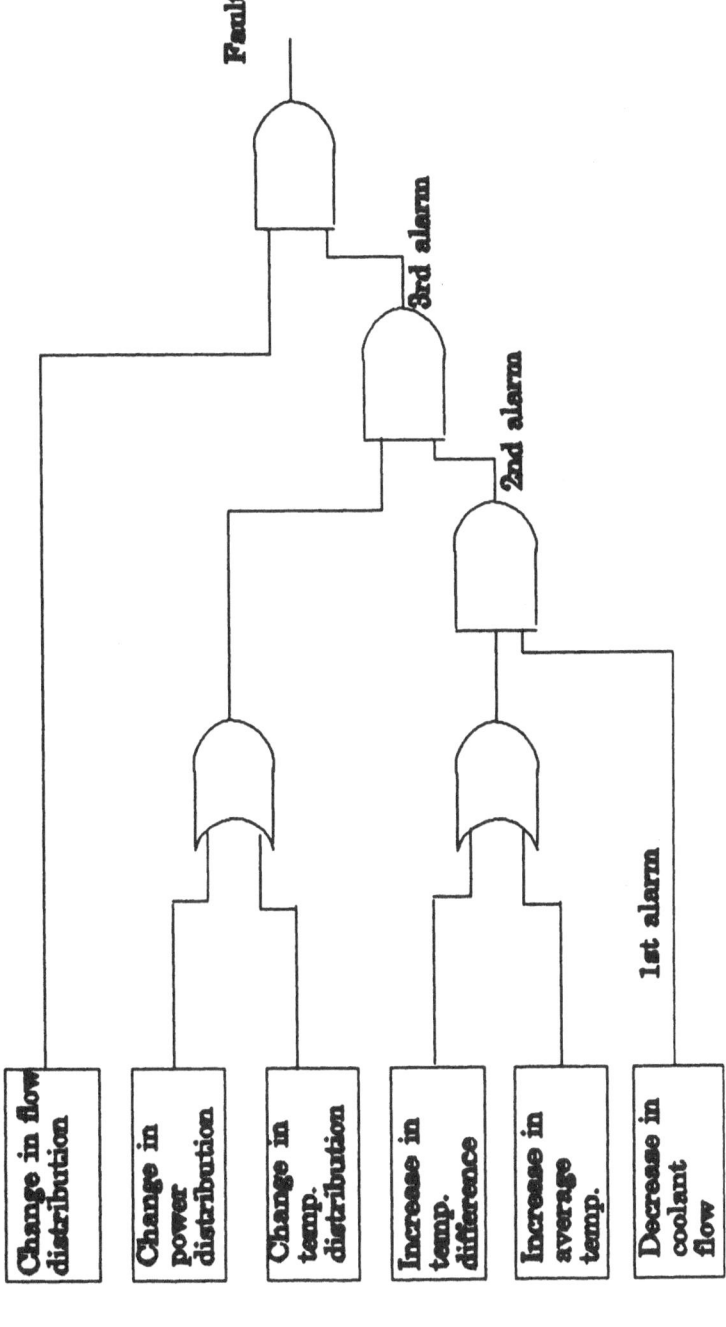

Figure 2: Fault tree for a partial loss of flow due to a blockage in the fuel assembly

# 5   Threshold Logic Approach for Fault Diagnosis

## 5.1   Threshold Logic Unit

A threshold logic unit is referred to a processing element with a step threshold function, so the output of the processing element is discrete, taking on values 1 or 0. These values can be used to solve logic problems where 1 corresponds to a fault condition while 0 corresponds to a normal condition.

The single threshold logic unit first initializes the connection weights and the threshold value to small random values and then computes a weighted sum of the input elements, adds a threshold $\Theta$ and passes the result through a step function such that the output Y is either 1 or 0 [7,16].

## 5.2   Threshold Logic Approach for Fault Diagnosis

New network models will be designed to assist the nuclear reactor operator in identifying the different accidents and alarms that might occur in the reactor. The approach is based on designing a network for each logic tree of the faults and alarms that might happen in the reactor. A threshold logic algorithm is applied to train the networks where the training algorithm involves calculating the output for the processing elements in the first layer, computing the error by comparing the calculated output with the desired value and adjusting the weight and threshold and changing the output so that the error is 0 before processing to the next layer of processing elements.

The inputs for these networks are discrete values of 1 and 0, where 1 refers to the occurrence of the symptoms of the fault which arises as a result of the deviation in the state variables of the simulated LOFT model, while the 0 refers to the normal operation situation. The inputs for the network are found by monitoring the state variables for any increase or decrease from the normal operating limit and the appearance of any sensor alarms concerning the whole reactor plant.

The output of the network is either 1 or 0 where 1 refers to the occurrence of the fault or the alarm while 0 means normal operation.

## 5.3   Applying the Threshold Logic Algorithm

The idea of this approach is based on processing through the network starting from the first layer, where the output for each processing element is calculated until the output of the last processing element in the final layer is found. If this output is 1 then the fault or the alarm has occurred otherwise this type of fault will not arise.

To demonstrate the performance of this approach, an example that deals with the failure of a power control system to respond to boron increase in which the network is shown in figure 3 is considered. The fault tree for this accident is illustrated in figure 4. The symptoms for this type of accident include, an increase in boron concentration, temperature difference across the reactor core and reactor power, and a decrease in the average coolant temperature and pressuriser pressure. Assuming all these symptoms have occurred, the input set to the network is (1 1 1 1 1), where the one's are the input values for the processing elements in the input layer. The network consists of two processing elements in the hidden layer and the output of the first one is calculated according to the threshold logic algorithm as

$$1 \times 1 + 0.5 = 1.5$$

Since $1.5 > 1.0$ then the output of this processing element is 1 indicating the first alarm of failure in the power control system to respond to boron increase. While the output for the second element is

$$1 \times 1 + 1 \times 1 + 1 \times 1 + 1 \times 1 - 2.5 = 1.5$$

Thus the output is also taken as 1. The output of the processing element in the output layer is then calculated using the same procedure and is found to be equal to 1.5 thus assuring the occurrence of this type of fault.

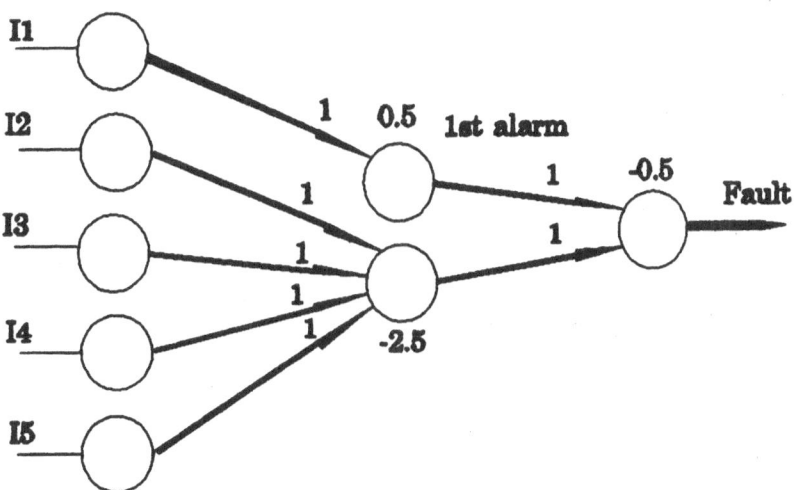

Figure 3: The network for failure in the power control system to respond to boron increase

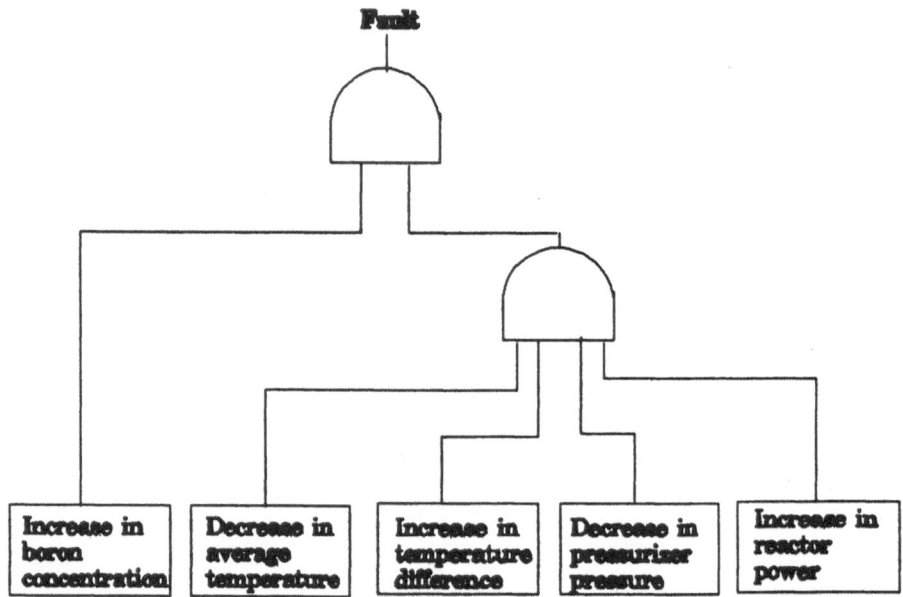

Figure 4: Fault tree for failure in the power control system to respond to boron increase

## 5.4 Network Result Using the Threshold Logic Approach

The computer program simulates the LOFT model, monitors the state variables for any deviation from normal operation, asks the user about the occurrence of other symptoms of different accidents, arranges and inputs the data to the networks and interprets the network output concerning the occurrence of any fault or alarm in the reactor. The program is designed to execute all these procedures every second.

The same hypothetical accident used to demonstrate the ability of the back propagation algorithm is used here to examine the ability of this system to assist the nuclear reactor operator in diagnosing accidents. The accident starts by a drop in the value of the coolant flow at a time greater than 2 sec. The neural network model for this type of accident is illustrated in figure 5.

Table 3 illustrates the output of the program where at time 0 sec the system indicates normal operation of the reactor. At time 3 sec it alarms the user for the partial loss of flow and at time 5 sec the system assures the occurrence of the first, second, and third alarms of partial loss of flow as well as the occurrence of the accident.

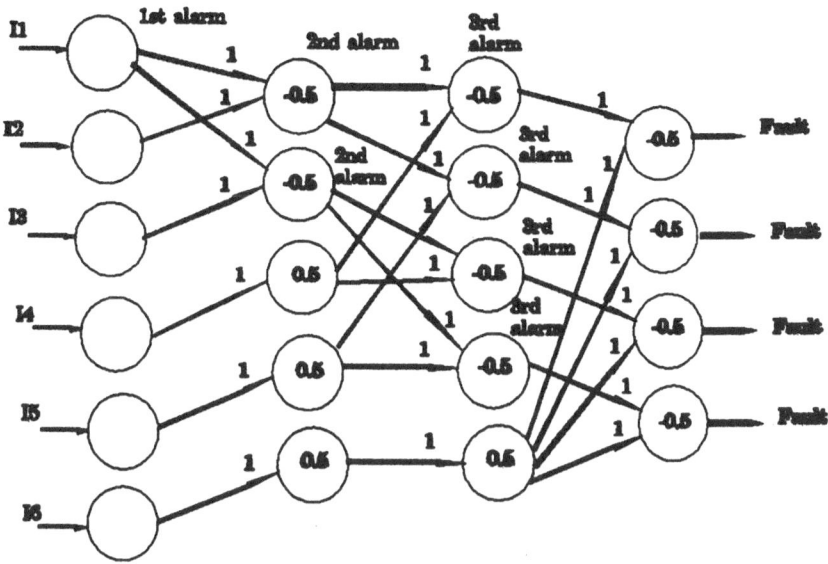

Figure 5: Network for partial loss of flow due to blockage in fuel assembly

| Time | Faults and alarms |
|------|-------------------|
| 0.0 | Nothing wrong with the reactor |
| 1.0 | Nothing wrong with the reactor |
| 2.0 | Nothing wrong with the reactor |
| 3.0 | First alarm for blockage in fuel assembly |
| 4.0 | First alarm for blockage in fuel assembly |
| 5.0 | First alarm for blockage in fuel assembly |
|  | Second alarm for blockage in fuel assembly |
|  | Third alarm for blockage in fuel assembly |
|  | Failure due to blockage in fuel assembly |

Table 3: The different faults and alarms with respect to time

# 6 Conclusion

The ability of the neural network to diagnose malfunctioning of the reactor is attractive and encouraging, and the back propagation algorithm worked very well in training the networks to identify the accidents. In all cases, the network output is very near to the desired output indicating that convergence has been achieved using the back propagation algorithm. This approach has been applied to fourteen different accidents and can be easily extended to include different models of faults and accidents that might occur in the reactor plant. The results of the threshold logic approach are also very good, and the approach is simple and easy to apply. The system can be used to represent any type of logic tree and to identify and diagnose any alarm or accident that might occur.

# References

[1] Kohonen, T. (1988). *An Introduction to Neural Computing.* Neural Networks, Vol. 1, pp. 3-16.

[2] Roth, M. (1988). *Neural-Networks Technology and It's Applications.* Johns Hopkins APL Technical Digest, Vol. 9, No. 3, pp. 242-253.

[3] Suddarth, S., Sutton, S. and Holden, A. (1988). *A Symbolic Neural Method for Solving Control Problems.* IEEE International Conference on Neural Networks, San Diego, California, Vol. 1, pp. I516-I523.

[4] Bhat, N. and McAvoy, T. (1989). *Use of Neural Nets for Dynamic Modeling and Control of Chemical Process Systems.* Proceedings of the American Control Conference, Pittsburgh, PA, Vol. 2, pp. 1342-1347.

[5] Hoskins, J. and Himmelblau, P. (1988). *Artificial Neural Network Models of Knowledge Representation in Chemical Engineering.* Comput. Chem. Engng., Vol. 12, No. 9/10, pp. 881-890.

[6] Saito, K. and Nakano, R. (1988). *Medical Diagnostic Expert System Based on PDP Model.* IEEE International Conference on Neural Networks, San Diego, California, Vol. 1, pp. I255-I262.

[7] Gallant, S. (1988). *Connectionist Expert Systems.* Communication of the ACM, Vol. 31, No. 2, pp. 152-169.

[8] Uhrig, R. and Guo, Z. (1989). *Use of Neural Networks to Identify Operating Conditions in Nuclear Power Plants.* Proc. SPIE Int. Soc. Opt. Eng., Vol. 1095, No. 2, pp. 851-856.

[9] Karna, K. and Breen, D. (1989). *An Artificial Neural Networks Tutorial: Part1-Basics.* Neural Networks, Vol. 1, No. 1, pp. 4-22.

[10] Simpson, P. (1990). *Artificial Neural Systems Foundations, Paradigms, Applications and Implementations.* Pergamon Press.

[11] McClelland, J. and Rumelhart, D. (1989). *Explorations in Parallel Distributed Processing. A Handbook of Models, Programs, and Exercises.* MIT Press.

[12] Rumelhart, D. and McClelland, J. (Eds.) (1986). *Parallel Distributed Processing: Explorations in the Microstructure of Cognition, Vol. 1: Foundations.* MIT Press.

[13] Werbos, P. and Park, C. (1988). *Back Propagation: Past and Future.* IEEE International Conference on Neural Networks, San Diego, California, Vol. 1, pp. I343-I353.

[14] Tylee, J. (1980). *Low Order Model of the Loss Of Fluid Test (LOFT) Reactor Plant for Use in Kalman Filter Based Optimal Simulation.* Proceedings of the 4th Power Plant Dynamics, Control and Testing Symposium, Gatlinburg, pp. 1-31.

[15] Tylee, J. (1980). *Low Order Model of the Loss Of Fluid Test (LOFT) Reactor Plant for Use in Kalman Filter Based Optimal Estimators.* EGG-2006, EG & G Idaho Falls, Idaho.

[16] Lippman, R. (1987). *An Introduction to Computing with Neural Nets.* IEEE ASSP Magazine, Vol. 4, No. 2,pp. 4-22.

# A Neural Network Approach to Tokamak Equilibrium Control

Chris Bishop

AEA Technology, Harwell Laboratory,
Oxfordshire OX11 0RA


Peter Cox, Paul Haynes,
Colin Roach, Mike Smith,
Tom Todd, and David Trotman

AEA Technology, Culham Laboratory,
Oxfordshire OX14 3DB
(Euratom/UKAEA Fusion Association)

### Abstract

We exploit the properties of the multilayer perceptron to develop a neural network approach to the feedback control of plasma position and shape in a tokamak experiment. The requirements of large bandwidth and high precision have led us to develop a custom hybrid analogue–digital hardware implementation of the neural network using conventional components. It is planned to demonstrate a complete system on the COMPASS tokamak at Culham Laboratory.

## 1    Introduction

Real-time feedback control of the position and shape of the plasma in a tokamak experiment represents a complex, non-linear problem. Early tokamaks had plasmas whose poloidal cross-section was circular, for which feedback control is relatively straightforward. However, recent tokamaks, such as the COMPASS experiment at Culham Laboratory and the JET (Joint European Torus) experiment, as well as most next-generation designs, have cross-sections which are strongly shaped. The accurate generation of such plasmas and the real-time control of their shape and position (which is usually unstable) presents a demanding problem.

In this paper we describe a neural network approach to the fast determination of plasma position and shape in a tokamak, based on the multilayer perceptron. This can be used as the basis for a real-time feedback control system, and offers a number of significant advantages over currently available techniques. It will also prove useful as a diagnostic tool providing rapid post-shot analysis of the plasma evolution. In addition the outputs of the network could be used as inputs to other high speed diagnostic data processing systems which may themselves be neural network based [1]. The combined requirements of high bandwidth and high precision have led us to develop a custom hybrid hardware implementation using analogue signal processing, together with digitally stored synaptic weights. The use of conventional components allows the development time to be relatively short. Prototype hardware modules have already been tested and it is intended to demonstrate the use of a complete system for feedback control on the COMPASS tokamak at Culham Laboratory.

In the next section we give a brief overview of the tokamak magnetic confinement system, and in Section 3 we describe the problem of plasma feedback control in a tokamak. The network architecture and training methods are described in Section 4, and in Section 5 results from software simulations of the network are presented. The hardware implementation of the network is described in Section 6, and some brief conclusions are given in Section 7.

# 2   Tokamak Overview

The tokamak is the principal experimental device for research into the magnetic confinement approach to controlled fusion. It consists of a toroidal vacuum vessel containing a small quantity of (isotopes of) hydrogen which is ionised and heated by a large toroidal electric current. This current is induced by transformer action using a time varying magnetic field which is itself generated by currents flowing through external coils. The plasma current also generates a magnetic field which helps to confine the plasma and thereby allow its temperature to be raised to very high values ($\simeq 10^8$ K). A strong toroidal magnetic field, generated by large toroidal-field coils, serves to stabilise the plasma. Additional magnetic fields, generated by currents flowing through poloidal-field coils, control the plasma shape and position.

Figure 1 is a cutaway drawing of the COMPASS tokamak showing the toroidal vacuum vessel as well as the support structure and a number of the control coils. Note that the cross-section of the vacuum vessel has a D-shape, unlike most earlier tokamak designs which were constructed with circular cross-section vessels. This D-shaped vacuum vessel allows non-circular cross-section plasmas to be generated. These are found to have improved energy confinement times and pressure limits, allowing significantly higher densities and temperatures to be achieved.

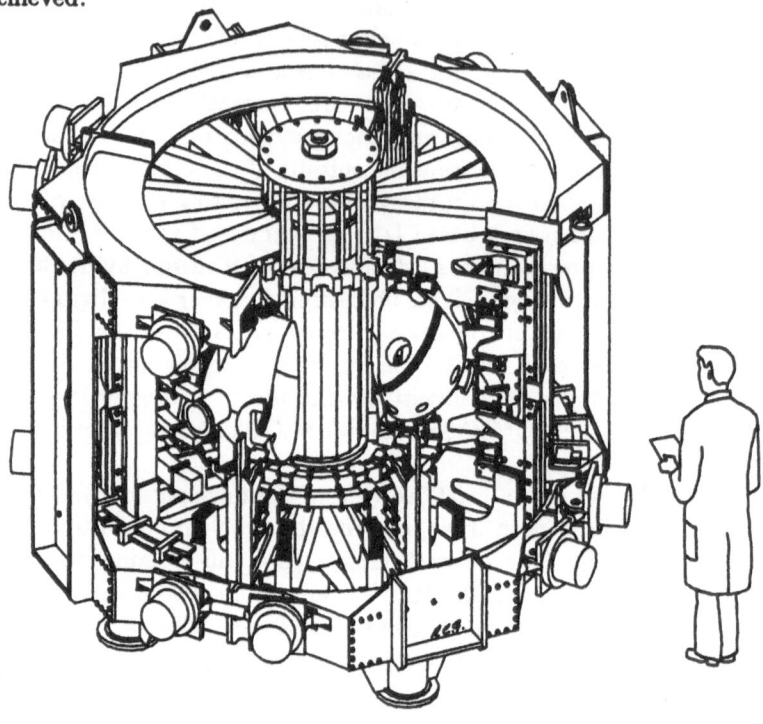

Figure 1

Some examples of the kinds of plasma shapes which can be produced in COM-PASS are shown in Figure 2. The Joint European Torus (JET) experiment (currently the world's largest tokamak), situated adjacent to Culham Labora-tory, as well as nearly all designs for future tokamak experiments, also have D-shaped vacuum vessels.

# 3    Tokamak Plasma Feedback Control

While non-circular cross-section plasmas offer a number of physics advantages over circular plasmas, they also present a number of additional problems. In particular, such plasmas are more difficult to produce and to control accurately. Currents through several control coils must be adjusted in order to generate the desired shapes. Furthermore, during a typical plasma pulse, the shape must evolve, usually from some initial near-circular shape. Due to uncertainties in the precise current and pressure distributions within the plasma, the most accurate approach to plasma control is to make real-time measurements of the position and shape of the boundary, and to use error feedback to adjust the currents in the control coils.

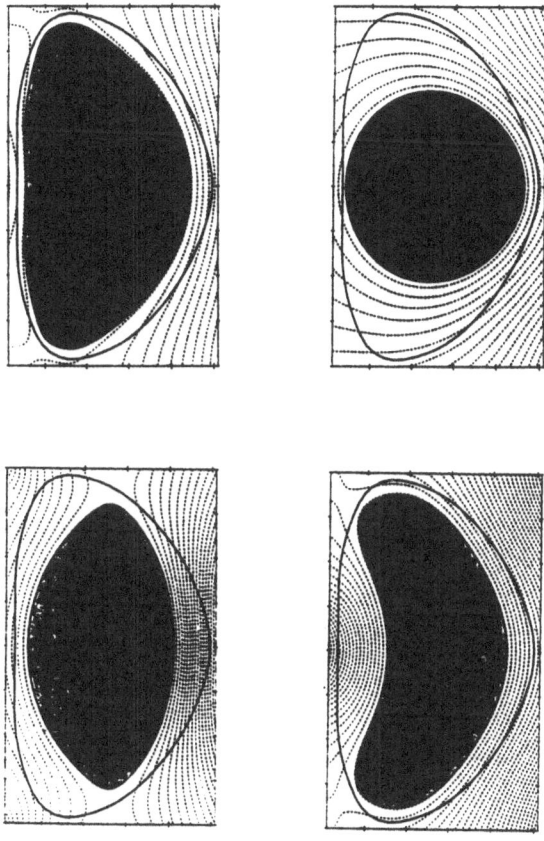

Figure 2

Equilibrium configurations of the plasma are described by solutions of the Grad-Shafranov equation, given by

$$R\frac{\partial}{\partial R}\left(\frac{1}{R}\frac{\partial\Psi}{\partial R}\right) + \frac{\partial^2\Psi}{\partial Z^2} = -RJ(\Psi, R) \qquad (1)$$

where $\Psi(R, Z)$ is the poloidal flux, so that the magnetic field lines are constrained to lie on contours of constant $\Psi$ (these contours are shown by the dashed curves in Figure 2). $R$ and $Z$ are respectively the radial and vertical coordinates, such that the $Z$-axis is the axis of symmetry of the tokamak.

The function $J(\Psi, R)$ specifies the toroidal plasma current density, which is determined by a variety of competing processes within the plasma such as impurity radiation and transport. Note that equation 1 assumes the tokamak to have rotational symmetry about the vertical axis. In the present context this approximation is more than adequate.

Due to the non-linear nature of the Grad-Shafranov equation a general analytic solution is not possible. However, for a given current density function $J(\Psi, R)$, the Grad-Shafranov equation can be solved by iterative numerical methods, with boundary conditions determined by currents flowing in the external control coils which surround the vacuum vessel. (On the tokamak itself it is changes in these currents which are used to alter the position and cross-sectional shape of the plasma). This is a standard technique for off-line analysis of the plasma, and is the method used to generate the dataset described in the next section, but it is computationally very intensive and is therefore unsuitable for real-time feedback control purposes.

For real-time control it is necessary to have a fast (typically $\leq 100\mu$sec.) determination of the plasma boundary shape. This information can be extracted from a variety of diagnostic systems, the most important being local magnetic measurements taken at a number of points around the perimeter of the vacuum vessel. Other diagnostic information can also be used, for instance information about the spatial profiles of density, temperature and poloidal magnetic field (which together help to determine the current profile function $J$) obtained from laser diagnostics. In the work reported here only magnetic signals have been considered, although it is intended to extend this to include other diagnostic information in the future.

The basic problem which has to be addressed, therefore, is to find a direct mapping from the measured diagnostic signals onto the values of a set of geometrical parameters describing the plasma boundary. This can be done by generating a large database of solutions of the Grad-Shafranov equation to which a functional form can be fitted to represent the required mapping. The mapping can then be implemented in suitable hardware and can run in real time. Conventional approaches have used various functional forms:

1. Dimensionality reduction using principal component analysis followed by a quadratic fit [2].

2. Ad-hoc hand-selected functional forms [6].

3. A piecewise-linear approximation. This requires that the hardware implementation allows the matrix elements of the linear mapping to be switched rapidly many times during the plasma pulse as the plasma evolves from one region of parameter space to another [3].

Approaches 1 and 2 assume specific functional forms for the mapping, which are chosen in a largely arbitrary manner, and which need not be optimal. While approach 3 can in principal approximate any continuous mapping with arbitrary accuracy, it may require large numbers of distinct linear regions to cover a complete plasma pulse, and suffers from the disadvantage of having a control law which is switched in real-time, with no guarantee of continuity.

In this paper we propose an alternative approach to the problem of tokamak equilibrium feed-back control based on feedforward neural networks [5]. As in other approaches, a large database of numerical equilibria is generated. The functional fit, however, is produced using a multilayer perceptron. Since the multilayer perceptron is capable of generating a very large class of non-linear mappings [4], the need to switch parameters can be avoided, while no specific limitation on the functional form of the mapping is introduced. Figure 3 shows a block diagram of the control loop. Information from tokamak diagnostics (principally magnetic pick-up coils) is fed to the inputs of the neural network, which is trained to generate the required geometrical parameters (minor radius, elongation etc.) as outputs. These are then compared with the desired values of those parameters (which are typically pre-programmed as functions of time prior to the plasma pulse) and the differences sent via standard linear controllers to the feedback amplifiers.

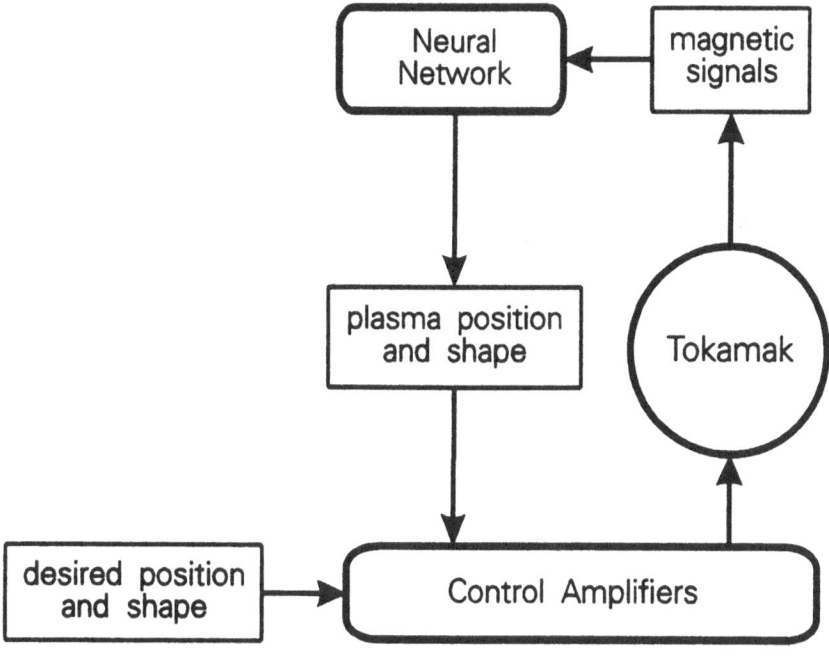

Figure 3

## 4   Dataset and Network Architecture

The dataset for training and testing the network was generated using a free boundary equilibrium code, and presently consists of approximately 1,000 equilibria (given by solutions of equation 1) spanning the wide range of plasma positions and shapes available in COMPASS. Since the shape of the plasma boundary is not strongly sensitive to the precise form of the current density function $J$, we have considered the specific form

$$J(\Psi, R) = b\left\{\beta R + \frac{(1-\beta)}{R}\right\}(1 - \Psi^{\alpha_1})^{\alpha_2} \qquad (2)$$

where $b$ is a constant, $\beta$ controls the ratio of plasma pressure to magnetic field energy density, and the parameters $\alpha_1$ and $\alpha_2$ can be varied in the range 1–6 to generate a variety of current profiles. For a large class of equilbria, the plasma boundary can be reasonably well represented by the parameterisation

$$\begin{aligned} R &= R_0 + a\cos(\theta + \delta\sin\theta) \\ Z &= Z_0 + a\kappa\sin\theta \end{aligned} \qquad (3)$$

where we have defined the following parameters

$R$    radial coordinate from the major axis of the torus,
$Z$    vertical coordinate from the torus midplane,
$R_0$   radial distance of the plasma centre from the major axis of the torus,
$Z_0$   vertical distance of the plasma centre from the torus midplane,
$\theta$    poloidal angle measured from the midplane,
$a$    minor radius measured in the plane $Z = Z_0$,
$\kappa$    elongation,
$\delta$    triangularity.

Each of the entries in the database has been fitted using the form in equation 3, so that the equilibria are labelled with the appropriate values of the shape parameters. Figure 4 shows scatter plots of pairs of boundary parameters for the whole dataset. These are seen to form distinct regions which correspond to the physical range of the tokamak operating space. Within these regions the parameter spaces are seen to be reasonably well populated. The plot of $R/a$ versus $a$, however, appears to be sparsely populated. This is due to the fact that the equilibria fall into distinct classes according to whether the plasma boundary makes contact with the vacuum vessel at the outer mid plane, the inner mid plane or some other point. Each configuration leads to a strong correlation between $R/a$ and $a$, as seen in the plot. Such correlations are also expected in experimentally achieved equilibria. The dataset was divided into two (randomly partitioned) halves for training and testing.

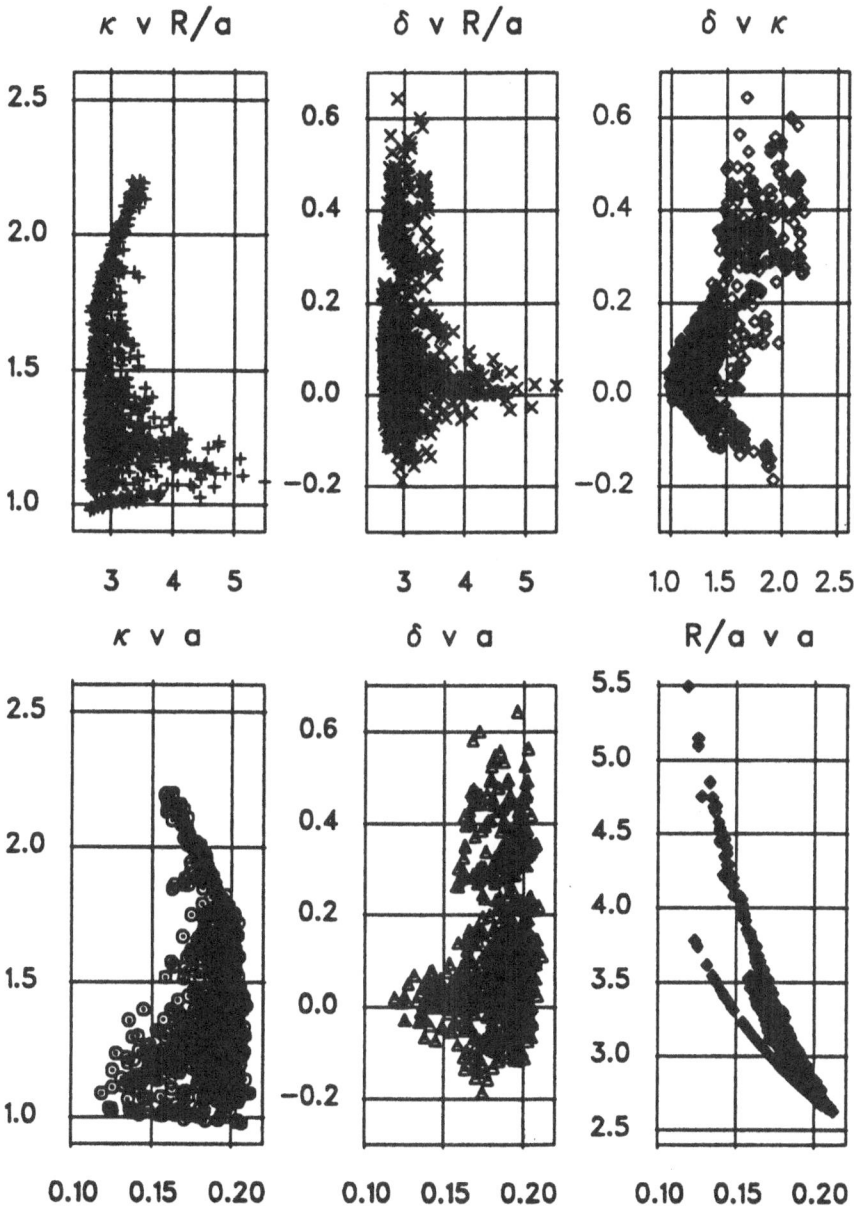

Figure 4

The results presented in this paper are based on a multilayer perceptron architecture with a single hidden layer. The units in the hidden layer compute a symmetric sigmoid function of the form

$$f(x) \equiv \frac{1}{1 + e^{-x}} - \frac{1}{2} \tag{4}$$

where $x$ is the input to the unit and $f(x)$ is the output. Units in the output layer are taken to be linear.

It is important to note that the transformation from magnetic signals to flux surface parameters involves an exact linear invariance. This follows from the fact that, if all of the currents are scaled by a constant factor, then the magnetic fields will be scaled by the same factor, and the geometry of the flux surfaces will be unchanged. It is important to take advantage of this prior knowledge and to build it into the network structure, rather than force the network to learn it by example. We therefore normalise the input vector to the network, and this must be reflected in the hardware design. The normalisation has an additional advantage; the input signals have a significant dynamic range (as the toroidal plasma current grows from a few kA to a few 100kA during the plasma pulse). Without this normalisation we would expect a loss of accuracy at low plasma currents due to the limited precision of the hardware implementation.

Error backpropagation is used to compute the derivatives of the mean-square-error with respect to the weights and thresholds in the network. A variety of training algorithms based on backpropagation have been explored. These included standard gradient descent with momentum and a number of enhanced gradient descent methods. However, the most powerful and robust methods were found to be conjugate gradients, and the BFGS (Broyden-Fletcher-Goldfarb-Shanno) memoryless quasi-Newton method. Results presented in this paper were obtained using conjugate gradients.

# 5   Simulation Results

Initial results were obtained on networks having four output units corresponding to minor radius $a$, major radius $R_0$, elongation $\kappa$ and triangularity $\delta$, these being parameters which are important for feedback control, and which must therefore be determined accurately. The results were also compared with those from the optimal linear mapping, that is the single linear transformation which minimises the same mean-square error as is used in the neural network training algorithm. This comparison provides some indication of the degree of non-linearity involved in the mapping. Although a more accurate mapping could be obtained using switched linear matrices, the switching involved would be undesirable in a control loop since there is no easy way to ensure continuity.

Networks with different numbers of input units and hidden units were trained, starting from randomly initialised weights. A single layer of hidden units was used with full interconnections between layers and no direct connections between input and output units. For a network with 26 inputs, for instance, the mean square error decreased as the number of hidden units was increased, saturating for about 12 hidden units. In all cases the mean-square error with respect to the test dataset was substantially smaller than for the equivalent optimal linear map (except when the number of hidden units was 3 or less, as might be expected since the data has then to be compressed into a space of fewer dimensions than the output space). For 12 hidden units, and 26 inputs, the normalised error for the test data was 0.14 as compared with 0.25 for the linear mapping. This difference is expected to become more marked as the range of equilibria in the dataset is increased and the required mapping becomes more non-linear.

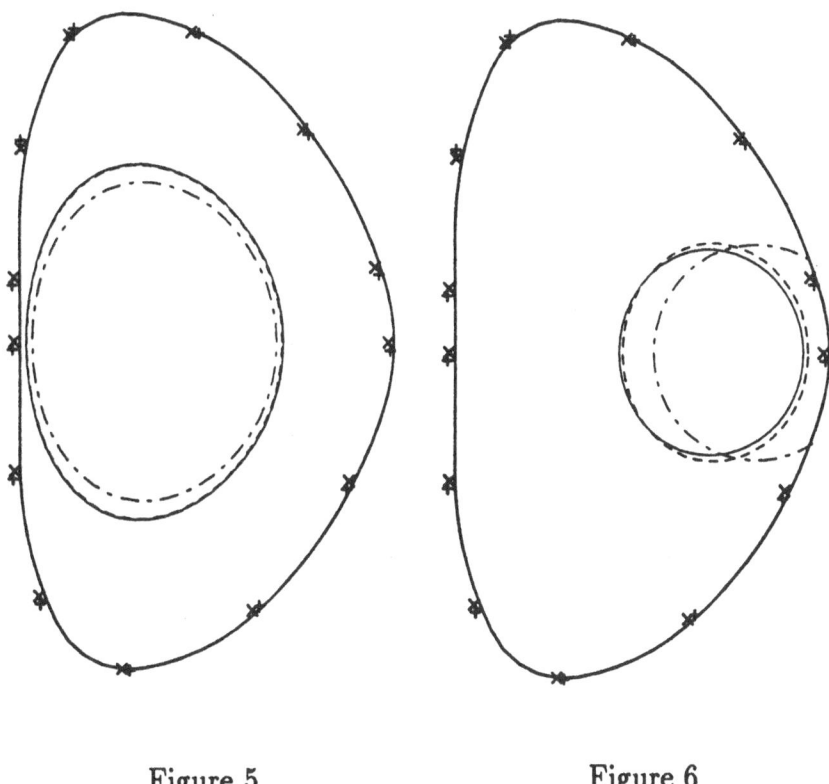

Figure 5                    Figure 6

Figure 5 shows the cross-section of the COMPASS tokamak vacuum vessel together with the magnetic pickup coil locations (indicated by the crosses) used to obtain these results. Also shown are representative plasma boundaries as generated by the numerical solution of equation 1 (solid curve), and as predicted by the neural network (dashed curve) and by the optimal linear mapping (dot-dash curve), for an example equilibrium from the test dataset. In this case, the boundary predicted by the neural network is barely distinguishable from that obtained from the Grad-Shafranov equation. Figure 6 shows an extreme example from the boundary of the test dataset. In this case the linear mapping has produced a pathological solution in which the plasma boundary intersects the wall of the vacuum vessel. For this example the network solution shows significant error, although it is substantially smaller than that arising from the linear mapping.

Figure 7 shows plots of the network predictions for various parameters versus the corresponding values from the database. Similar plots for the optimal linear map predictions versus the database values are given in Figure 8. (Note that these particular results were obtained using an earlier form of the database, spanning a slightly smaller range of parameters than depicted in Figure 4). Comparison of these two figures shows the significantly poorer predictive capability of the linear mapping, and indicates that significant non-linearities are indeed required in the mapping.

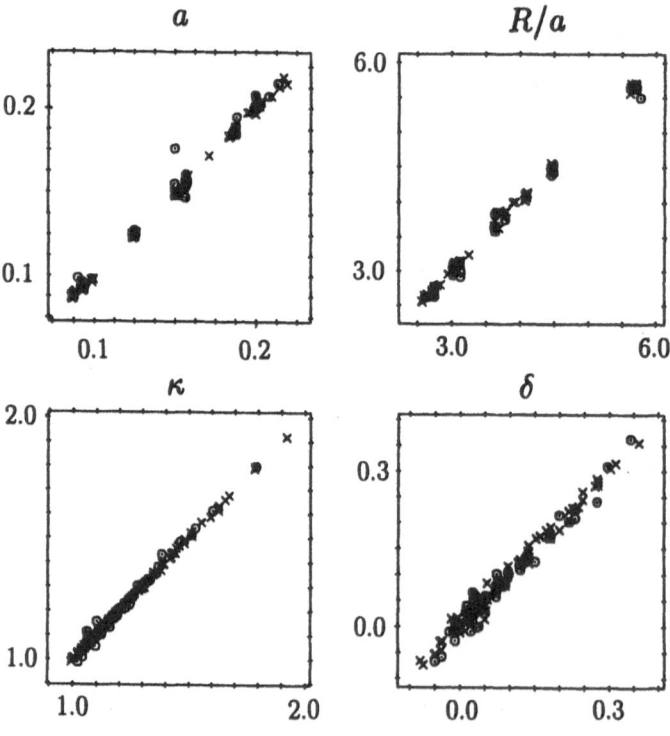

Figure 7

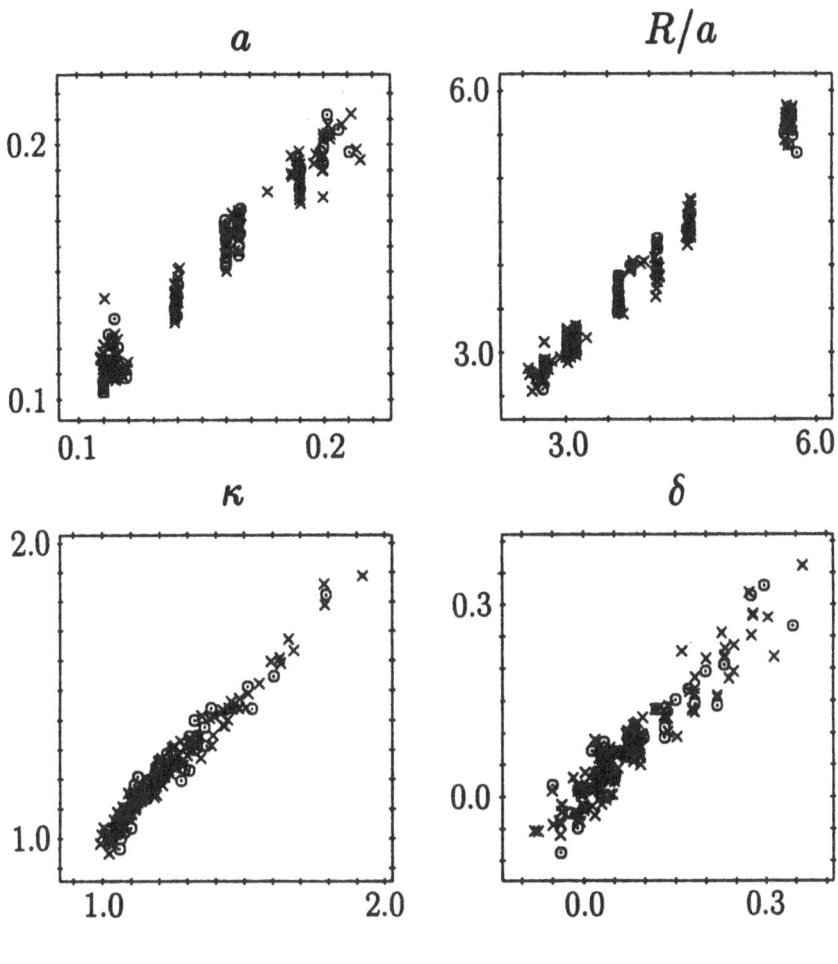

Figure 8

# 6    Hardware Implementation

The hardware implementation of the neural network must satisfy a number of requirements. A bandwidth of $\geq 10$ kHz is needed to cope with the fast timescales of the plasma evolution, an output precision of at least 8 bits is required in order to achieve sufficiently accurate control of the plasma boundary, and the system must be able to cope with the large dynamic range of the input signals. It is also desirable, for development work, to have a system in which the network architecture can be readily changed in order to explore different numbers of inputs, different numbers of hidden units, the effects of a second hidden layer, and so on. Finally, the hardware must be developed on a relatively short timescale in order to fit into the desired COMPASS schedule.

The requirement for 8 bit precision suggests that the synaptic weights should be stored to significantly higher precision than this. Prelimary simulations of the hardware network indicate that in feedforward mode there is little degradation in error as the precision of the weights is reduced until about 10 bits precision is reached. The error then starts to increase, becoming unacceptable for precisions less than about 8 bits.

We have chosen to develop a hybrid system using analogue signals to achieve high bandwidth and to avoid having to transfer signals between digital and analogue representations. (Both the diagnostic signals and the inputs to the feedback amplifiers are analogue). The synaptic weights are produced using 12 bit multiplying DACs (digital-to-analogue converters) which can be configured to allow four quadrant multiplication of analogue signals by a digitally stored number. (Note that the DACs are being used here as digitally controlled attenuators, and not in their usual role to convert digital signals into analogue signals). Thresholds are easily produced using an extra weight with a DC bias applied to the input.

There are many ways to produce a sigmoidal non-linearity and we have opted for a solution using two transistors configured as a long-tailed-pair. This arrangement generates a transfer characteristic of the form given in equation 1. The actual characteristic obtained from measurements on a prototype of the sigmoid is shown in Figure 9. Noise levels and response times are found to be well within the required limits. Corrective circuitry is also included to compensate for the effects of temperature variation.

A VME based modular construction has been chosen to allow for flexibility in changing the network architecture, and to allow defective segments to be replaced rapidly and so minimise down-time on the tokamak. Separate cards are used for the synaptic weights and for the sigmoids, and the network architecture is set up by inserting the appropriate set of cards into the rack and making the necessary interconnections on the front panels. The synaptic weights are downloaded (prior to the plasma pulse) via the VME backplane from a central control computer, using an addressing system to label the individual weights. The system will also have the facility for on-line weight changes to be made, allowing for possible future experiments with adaptive networks. In addition, extensive monitoring facilities are provided to allow internal signals to be digitised and stored for post-shot analysis.

Due to the significant dynamic range of the measured signals, as discussed earlier, an analogue hardware implementation of the input vector normalisation has also been developed. Buffering and power supplies complete the system, which is compact and has modest power requirements. The use of this hardware implementation for real-time feedback control of a tokamak plasma will be demonstrated on the COMPASS experiment at Culham Laboratory.

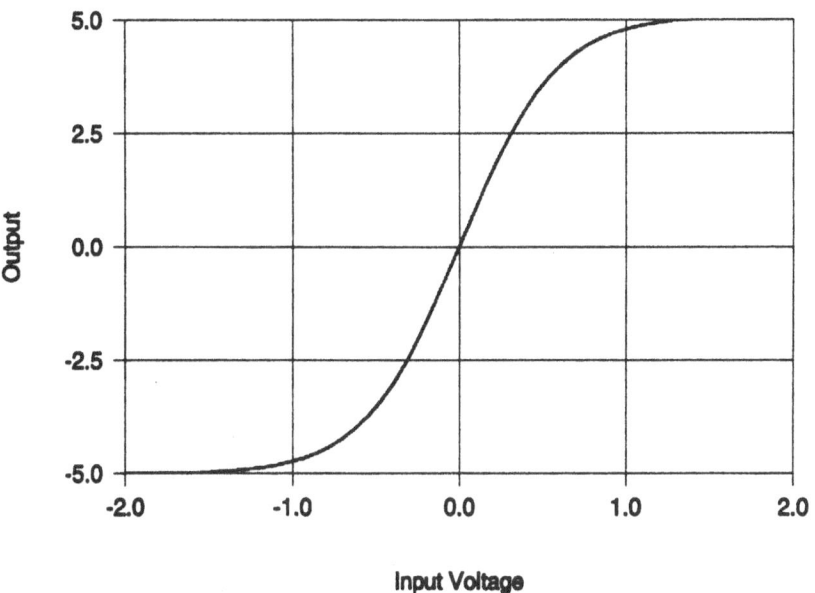

Figure 9

# 7 Conclusions

We have shown that the multilayer perceptron offers an alternative approach to the problem of real-time feedback control of the plasma in a tokamak experiment. Unlike many conventional approaches it does not assume a specific functional form for the mapping needed to determine the plasma shape, nor does it involve switching of the mapping in real-time. Simulation results demonstrate that the technique is capable of achieving the desired accuracy over a wide range of equilibrium configurations.

Custom neural network hardware has been designed which combines analogue signal paths with 12 bit synaptic weights to give high bandwidth together with high precision. This hardware system is readily extendable and can be easily reconfigured in a wide variety of architectures.

# References

[1] Bishop C M, Strachan I G D, O'Rourke J, Maddison G P and Thomas P R (1992) *Reconstruction of Tokamak Density Profiles using Feedforward Networks* Neural Computing and Applications **1** No 1.

[2] Braams B J, Jilge W and Lackner K (1986) *Fast Determination of Plasma Parameters Through Function Parameterisation*, Nuclear Fusion **26** No 6, 699.

[3] Hofmann F and Tonetti G (1988) *Fast Identification of Plasma Boundary and X-Points in Elongated Tokamaks*, Nuclear Fusion **28** No 3, 519.

[4] Hornik K, Stinchcombe M and White H (1989) *Multilayer Feedforward Networks are Universal Approximators*, Neural Networks **2** No 5, 359.

[5] Lister J B and Schnurrenberger H (1991) *Fast Non-Linear Extraction of Plasma Parameters Using a Neural Network Mapping*, Nuclear Fusion **31** no 7, 1291.

[6] Osborne T H, Fukumoto H, Hosogane N et al (1986) *Plasma Shape and Position Control on DIIID*, Bull. Am. Phys. Soc. (1986) **31** No 9, 1502.

# Neural Network Control of Robot Arm Tracking Movements

Dean Shumsheruddin

School of Computer Science, University of Birmingham
Birmingham, UK

### Abstract

This paper describes a neural network based controller for tracking moving objects with a two-joint robot arm. The neural network consists of a single layer neural map containing pairs of cells, and two output neurons. The inputs to the network are the position of the hand, and the position and velocity of the object relative to the hand. The outputs from the network are the torques to be applied to the two joints.

The network learns the mapping between joint torques and hand movements by making random movements of the arm. During learning, the delta rule is used to adjust the weights of the connections between the neural map and the output neurons.

The model has been tested by computer simulation of a system consisting of the control network, a two joint arm with inertia and damping, and a randomly moving object. In the simulation, the neural network controller tracks objects with position and velocity errors of the order of 0.5 percent of the range of movement of the arm. This technique can be applied to a wide range of tracking control problems.

## 1 Introduction

Research on neural networks for motor control has two main objectives. The first is to understand how animals' movements are controlled by their nervous systems. Neural network models have shown how animals can learn accurate sensory-motor coordination despite unforseen changes in the dimensions of their bodies and strengths of their muscles. The second aim is to build controllers for practically useful robots. The main advantages of neural network controllers over conventional control systems are their efficiency and adaptability [1].

Most investigations of neural network models for motor control have studied the control of human or robot arm movements, and considerable progress has been made in this area. Bullock and Grossberg [2] have developed a neural network model of human arm movement control, called VITE, which accurately reproduces the bell-shaped velocity profiles of human arm movements. Gaudiano and Grossberg [3] have extended the system to produce the AVITE model for adaptive control of arm movement trajectories during visually guided reaching.

Kuperstein and his colleagues [4,5] have developed a neural controller called INFANT. This controller can learn to grasp objects, detected by a stereo pair of TV cameras, in three-dimensional space with a multi-joint robot arm. It can also learn to position unforseen payloads at the end of a single-joint arm with accurate and stable movements. Kawato et al [6] have developed a neural network controller for a two-joint robot arm. This can generate the torques required to produce smooth movements from one position to another, avoiding obstacles or passing through via-points. It uses the minimum torque change criterion of trajectory smoothness. Bruwer and Cruse [7] have developed an elegant neural network controller for programming the movement of redundant manipulators with multiple joints, although it outputs joint angles rather than torques.

Most of the work on the control of robot arm movement has concentrated on simple straight line movements from point to point. Tracking and catching moving objects is a more complex problem, although some progress has been made in this area. Miller [8] has developed a CMAC (cerebellar model arithmetic computer) based neural network controller for following objects moving in a straight line, with constant velocity, along a conveyor belt. This system takes input from a TV camera mounted on the end of the robot arm and outputs joint velocity control signals.

Tracking randomly moving objects in two dimensions with a robot arm poses more difficult problems. This paper presents a theoretical description and a computer simulation of a neural map based controller for tracking moving objects in a two-dimensional plane with a two-joint robot arm.

## 2    The Neural Network Controller

The controller performs the task of tracking a moving object with a two-joint robot arm as shown in Figure 1. The arm and the object are both confined to the two-dimensional plane, with coordinates $x$ and $y$. The object moves around randomly within this plane. The shoulder of the arm is fixed at the origin of the plane and the hand is free to move within the rectangular workspace shown in the figure. The links of the arm are assumed to have masses and moments of inertia. $\theta$ and $\phi$ are the shoulder and elbow joint angles. The arm is moved by actuators which can apply variable torques to the two joints. The joints are assumed to have fixed damping factors.

The controller has to specify the torques to be applied to the joints in order for the hand to track the object within the workspace. The inputs to the controller during tracking are the position of the hand in the workspace, and the position and velocity of the object relative to the hand. These quantities are all vectors with real-valued $x$ and $y$ components, continuously changing in time. The inputs to the controller during learning are the position and acceleration of the hand in the workspace. It is assumed that these inputs are provided by the robot arm's sensory systems.

The controller assumes very little about the system. It needs to be programmed with the coordinates of the workspace and the range of activation of the joint actuators. However it does not need to be given the lengths, masses

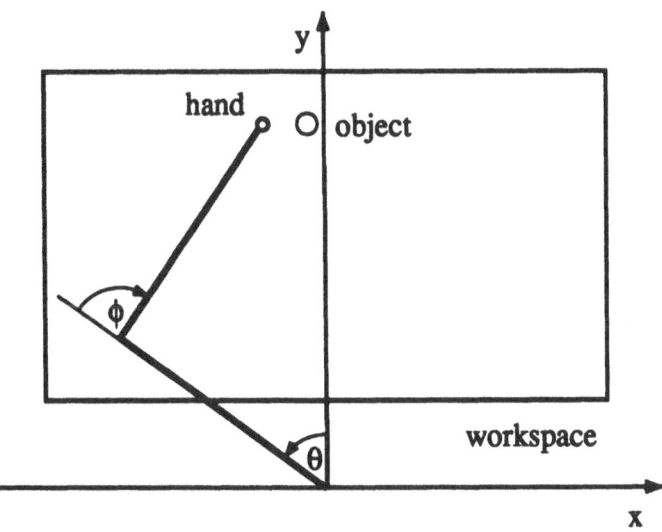

Figure 1: The Robot Arm

or moments of inertia of the arm segments, or the damping coefficients of the joints.

The structure of the neural network controller is shown in Figure 2. It consists of a two-dimensional neural map, of a similar type to that used in Kuperstien's INFANT system [4], and two torque output nodes, $T_1$ and $T_2$. The neural map consists of a rectangular array of pairs of neurons. Each pair of neurons represents the corresponding position in the workspace. During operation, inputs are gated into a circular region of the map centered on the position representing the current location of the hand. This is represented by the grey circle in Figure 2.

All the nodes in the map are connected to both torque output nodes. The strengths of these connections are set during the learning phase. Figure 3 shows the connections of a single pair of nodes in the neural map. One neuron in the pair processes the $x$ components of the input signals and the other processes the $y$ components. $\Delta x$ and $\Delta y$ are input signals representing the $x$ and $y$ components of the position of the object relative to the hand. $\Delta x_v$ and $\Delta y_v$ are input signals representing the $x$ and $y$ components of the velocity of the object relative to the hand. The weights $W_1$ and $W_2$ are constant. The weights between the map nodes and the output nodes ($W_a, W_b, W_c$ and $W_d$) are set during the learning phase. The torque output nodes, T1 and T2, are connected to the joint actuators at the shoulder and elbow joints respectively.

During tracking, the object position and velocity inputs are gated into the circular region of the map centered on the position representing the current location of the hand. The map nodes in this region calculate the $x$ and $y$

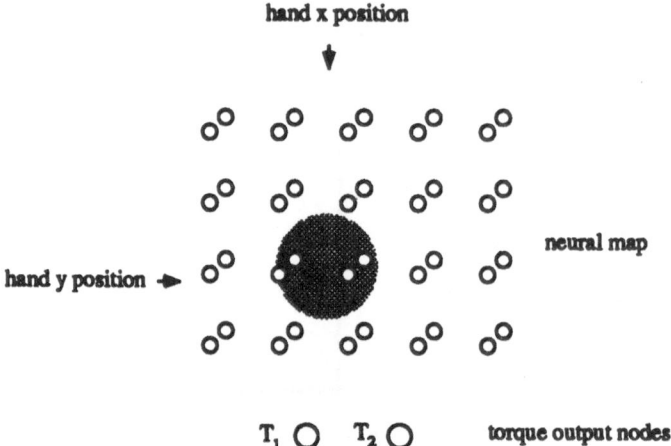

hand x position

hand y position →

neural map

$T_1$ ○   $T_2$ ○    torque output nodes

Figure 2: The Neural Network

components of the hand acceleration required to track the object. The output nodes calculate the joint torques required to produce this acceleration.

The object position and velocity signals are gated into the neural map by being multiplied by a gating function. The gating function for the pair of nodes at location $i,j$ in the map is given by the following formula:

$$G_{ij} = exp(-\lambda(\delta x^2 + \delta y^2))$$

where $\delta x$ and $\delta y$ are the $x$ and $y$ components of the difference between the position represented by node pair $ij$ and the current hand position, and $\lambda$ is a constant. This function has a value of one at the hand position and decays toward zero with increasing distance from the hand position. This gating function is based on that used in Kuperstien's INFANT system [4].

The model assumes that the hand acceleration required to track the object is given by:

$$a = k_1\Delta p - k_2\Delta v$$

where $\Delta p$ is the position of the object relative to the hand and $\Delta v$ is the velocity. $k_1$ and $k_2$ are constants, with values of 1.0 and 0.5 in the computer simulation. This simple function performs the task well and can easily be calculated by a pair of nodes. The gated input signals propagate, through the constant weights $W_1$ and $W_2$, into the map nodes. The activity of the $x$-component node of the pair $ij$ is given by the following formula:

$$A_{ijx} = G_{ij}\Delta x W_1 + G_{ij}\Delta x_v W_2$$

where $\Delta x$ is the $x$ component of the position of the object relative to the hand and $\Delta x_v$ is the $x$ component of its velocity relative to the hand. The activity of the $y$ node of the pair is calculated similarly.

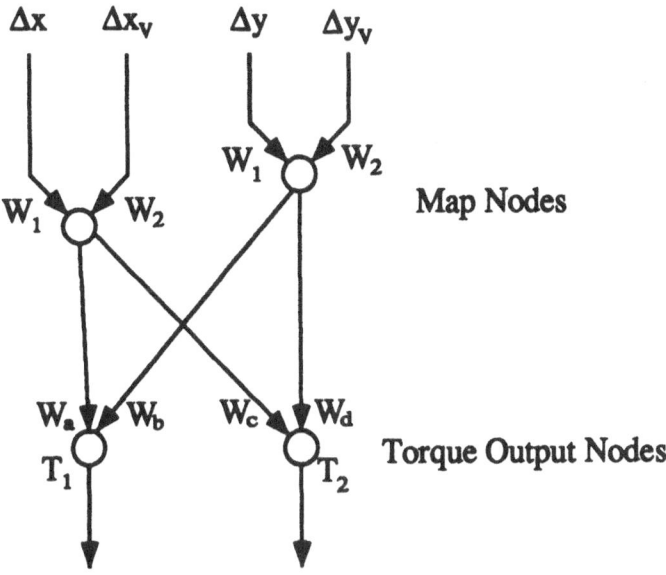

Figure 3: The connections of a single pair of map nodes

Activity of the map nodes propagates into the torque output nodes according to the following formula:

$$T_i = \sum_{j=1}^{N} A_j w_{ij}$$

where $T_i$ is the activity of the $i$th output node, $j$ ranges over all the N nodes in the map, and $w_{ij}$ is the weight of the link from node $j$ to node $i$.

The controller learns the mapping between hand acceleration and joint torques over the workspace during a learning phase of operation. It does this by setting the hand to random positions in the workspace and making random movements of the arm at each position. During a learning trial, the hand is first set to a random position in the workspace. Then random torques, within the range of operation of the actuators, are applied to the two joints. The $x$ and $y$ components of the resulting hand acceleration are gated into the pairs of map nodes. Activity is propagated through the network to produce two joint torque signals. These are compared with the actual torques $\tau_i$ that were applied, to produce an error signal for each output node:

$$E_i = \tau_i - T_i$$

Finally, the delta rule is used to correct the weights of the connections from all nodes in the map to the two output nodes:

$$\Delta w_{ij} = \eta E_i A_j$$

where $\eta$ is a constant learning rate.

# 3   Computer Simulation

A computer simulation was developed in order to test the neural network controller and investigate its performance on a number of tracking problems. The simulated system consisted of the neural network controller, a two-joint robot arm with inertia and damping, and a randomly moving object to be tracked by the arm.

The simulation was implemented in the C programming language on a Sun Microsystems Sparc workstation. The behaviour of the whole system was simulated with a time interval of one millisecond. All real-valued quantities in the simulation were represented by 64-bit floating point variables.

The neural map consisted of a 29 by 13 rectangular array of pairs of nodes, containing 754 nodes in total. The nodes represented an array of points in the workspace with a spacing of 5cm. The constant $\lambda$ was set to 500 to produce a gating function with a value of one half at a radius of 3.7 cm. The weights $W_1$ and $W_2$ were constant throughout the map with values of 10.0 and 5.0 respectively. The weights $W_a$ to $W_d$ were initialised with small random values and adjusted by the learning process described below.

The simulation of the arm was based on the mathematical model of the dynamics of a two-joint arm developed by Kalveram [9,10]. The model was simplified by assuming the arm and object moved in a horizontal plane and the effects of gravity could be ignored. The dynamics of the arm are described by two equations:

$$\text{Joint 1:} \quad A\ddot{\theta} + C\ddot{\phi} - D\dot{\phi}^2 - 2D\dot{\theta}\dot{\phi} + R_1\dot{\theta} = T_1$$
$$\text{Joint 2:} \quad B\ddot{\phi} + C\ddot{\theta} + D\dot{\theta} + R_2\dot{\phi} = T_2$$

where:

$$A = M_1 + M_2 + m_2 l_1^2 + l_1 a_2 m_2 cos\phi$$
$$B = M_2$$
$$C = M_2 + l_1 a_2 m_2 cos\phi$$
$$D = l_1 a_2 m_2 sin\phi$$

where $\theta$, $\dot{\theta}$ and $\ddot{\theta}$ are the angle, angular velocity and acceleration at the shoulder joint. $\phi$, $\dot{\phi}$ and $\ddot{\phi}$ are the angle, angular velocity and acceleration at the elbow joint, relative to the upper arm. $M_1$, $m_1$, $l_1$ and $a_1$ are the moment of inertia, mass, length, and distance from the joint to the centre of mass, of the upper arm. $M_2$, $m_2$, $l_2$ and $a_2$ are the corresponding quantities for the forearm. $R_1$ and $R_2$ are the moments of damping for the two joints and $T_1$ and $T_2$ are the torques exerted by the joint actuators. $A$ and $B$ represent the moments of inertial resistance against active acceleration torques about the joints. $C$, $-D$, $D$ and $2D$ represent the dynamic coupling effects due to reactive acceleration, centrifugal and centripetal forces in the two-joint system.

In the simulation, the bottom left corner of the workspace was at $(-0.7\text{m}, 0.1\text{m})$ and the top right corner at $(0.7\text{m}, 0.7\text{m})$. Both arm segments were

0.5m long, had masses of 0.5kg and had centres of mass at their midpoints. The damping factors of both joints were 0.1 and the torque outputs of both actuators could vary between −1.0Nm and +1.0Nm.

The object was simulated as a fixed mass randomly changing its velocity every 500ms. The maximum velocity of the object was about half that of the hand, with the arm in a typical position.

In order to simulate tracking movements, the states of the network, arm and object were calculated every millisecond. During each time interval, the inputs to the network were set from the simulated positions and velocities of the object and hand. Then the values of the gating function and the gated inputs were calculated for each pair of nodes in the map. Next, the activities of all the map nodes were calculated and propagated to the torque output nodes. To simulate the movement of the arm, angular accelerations were calculated from the torques output by the network, using the two formulae above, and angular velocities and positions were calculated by numerical integration. Then the object's velocity was determined and its position calculated by numerical integration.

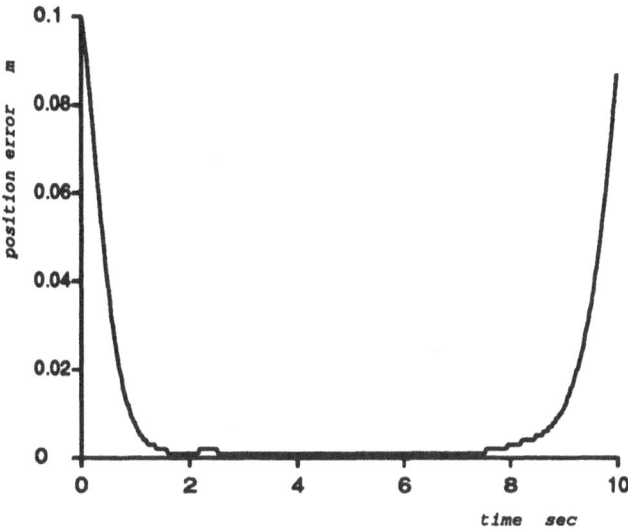

Figure 4: Position error over the course of a typical tracking movement

Figure 4 shows the distance between the hand and the object over the course of a typical tracking movement. The hand starts 0.1m away from the object, moves towards it over the first 2 seconds, and then tracks it for 4 seconds. After 8 seconds the object moves out of the workspace and the error increases.

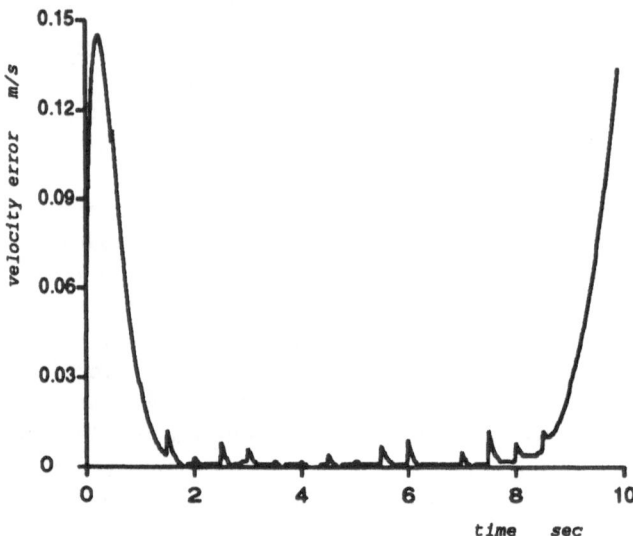

Figure 5: Velocity error over the course of a typical tracking movement

Figure 5 shows the difference in the velocities of the hand and object during the same movement. The small peaks in the velocity error graph are due to the random changes in the object's velocity every 500ms. The tracking errors were found to be minimal in the centre of the workspace with the elbow joint angle close to $\pi/2$. The errors were larger at the edges, and particularly in the corners of the workspace.

During learning, the hand was set to random positions in the workspace and random torques were applied to the two joints. The motion of the arm was simulated for three milliseconds and the hand acceleration measured. Then the gating function was calculated at each pair of nodes and the acceleration was multiplied by the gating function to set the activity of the nodes in the map. Activity was propagated to the output nodes and the errors were determined for each node. Finally the weights of all the connections from the neural map to output nodes were modified using the delta rule, as described above.

Figure 6 shows the mean error in the torques output by the network, for groups of 100 random movements, over the course of a typical learning phase. The optimum value for the learning rate $\eta$ was found to be about 0.01. Larger values of $\eta$ led to a failure of the weights to converge.

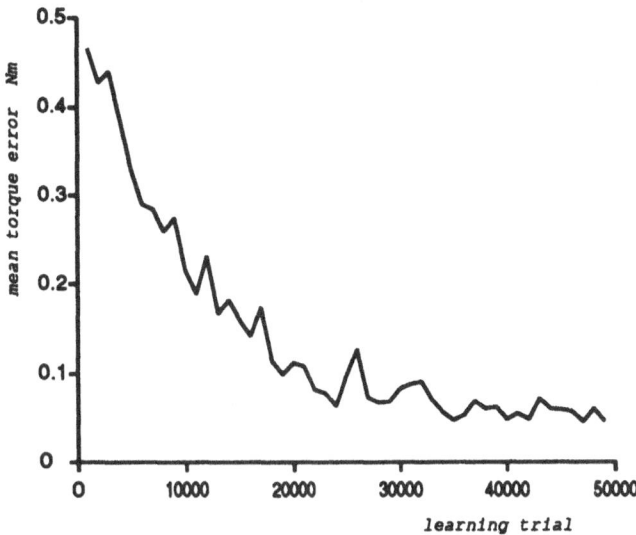

Figure 6: Mean torque output node error over the course of learning

# 4 Discussion

This paper has described a neural map based controller for tracking moving objects with a two-joint robot arm. It has also described the testing and evaluation of the controller's performance by means of a computer simulation. The results from the computer simulation clearly demonstrate that the controller can learn to track moving objects with considerable accuracy.

The principal advantages of neural network based controllers over conventional control systems are efficiency and flexibility. They also exhibit graceful degradation of performance with errors in input data or damage to the controller. Their principal disadvantage is the need for training.

Neural map based controllers have a number of advantages over alternative neural network control systems such as the CMAC based system developed by Miller [8]. Neural map based controllers are fast because they have a simple two-layer structure and can accept raw sensor signals as inputs and produce raw actuator control signals as output.

Neural map based controllers are flexible because they can learn optimal control of a system over the full range of its operating space. They can be used to control a system with any required degree of accuracy by increasing the density of the nodes in the neural map until the desired level of accuracy is achieved. The main disadvantages of neural map based controllers are the relatively large numbers of neurons and connections required, and the large number of training cycles needed for accurate performance.

The controller described here could be applied to a wide range of tracking control problems with relatively few modifications. For example, it could be used to control a TV camera with two degrees of freedom in a visual tracking system, such as that described by Aloimonos and Tsakiris [11].

Neural map based controllers can be applied to a very wide range of applications. A neural map can be used to represent the operating space of a complex system, such as the speed power-output space of an internal combustion engine. A neural map based controller can learn optimal control of such a system over the whole operating space.

The robot arm controller has a number of limitations. Perhaps the most serious is the need for a separate, and rather lengthy, learning phase of operation. A controller which could learn during normal tracking operation would be more useful as it could continuously adapt to changes in the controlled system due to factors such as component wear and minor faults. The system could be extended to control arms with more joints by simply adding more torque output nodes for the extra joint actuators. However, extending the system to work in three dimensions would be more difficult. In principle this could be done with a three dimensional map of triplets of nodes, but this would require a rather a large number of nodes.

Further work is in progress on extending the system to catch moving objects. It is also planned to develop a new version of the controller which can learn during normal tracking movements.

# References

[1] Werbos, P. J. (1991). An overview of neural networks for control. *IEEE Control Systems Magazine*, January 1991, 40-41.

[2] Bullock, D. and Grossberg, S. (1988). Neural dynamics of planned arm movements: Emergent invariants and speed-accuracy properties during trajectory formation. *Psychological Review* 95,49-90.

[3] Gaudiano, P. and Grossberg, S. (1991). Vector associative maps: Unsupervised real-time error-based learning and control of movement trajectories. *Neural Networks* 4, 147-183.

[4] Kuperstein, M. (1991). INFANT neural controller for adaptive sensory-motor coordination. *Neural Networks*, 4, 131-145.

[5] Kuperstein, M. and Wang, J. (1990). Neural controller for adaptive movements with unforseen payloads. *IEEE Transactions on Neural Networks*, 1, 137-142.

[6] Kawato, M., Maeda, Y., Uno, Y. and Suzuki, R. (1990). Trajectory formation of arm movement by cascade neural network model based on minimum torque change criterion. *Biological Cybernetics*, 62, 275-288.

[7] Bruwer, M. and Cruse, H. (1990). A network model for the control of the movement of a redundant manipulator. *Biological Cybernetics*, **62**, 549-555.

[8] Miller, W. (1989). Real-time application of neural networks for sensor-based control of robots with vision. *IEEE Transactions on Systems, Man and Cybernetics*, **19**, 825-831.

[9] Kalveram, K. T. (1991). Pattern generating and reflex-like processes controlling aiming movements in the presence of inertia, damping and gravity. *Biological Cybernetics*, **64**, 413-419.

[10] Kalveram, K. T. (1991). Controlling the dynamics of a two-jointed arm by central patterning and reflex-like processing. *Biological Cybernetics*, **65**, 65-71.

[11] Aloimonos, J. and Tsakiris, D. (1991). *Image and Vision Computing 9*, 235-251.

# The pRAM as a hardware-realisable neuron

*T G Clarkson*

Department of Electronic and Electrical Engineering,

King's College London, UK

### Abstract

The pRAM is a non-linear stochastic device with neuron-like behaviour. The pRAM generates an output in the form of a spike train. The spike train may be integrated over a variable number of time steps to control the amount of noise in the system. The pRAM is realisable in hardware and VLSI devices have been built. Learning algorithms have been developed for the pRAM which may be incorporated in hardware and the intrinsic noise in the pRAM has proved beneficial during training. A current application of a pRAM net is in real-time object recognition.

## 1  The pRAM

The pRAM is a non-linear stochastic device with neuron-like behaviour. The pRAM generates an output in the form of a spike train. The usefulness of these features can be demonstrated when compared to other neuron models.

The models of Figure 1 are neuron-like owing to their multiple inputs and single output. Inputs and outputs are binary signals. The McCulloch-Pitts (MCP) neuron [1] implements the connection weights between neurons using (typically) variable resistances which may be modified to establish a learnt behaviour. The output of the MCP neuron is a weighted sum of its inputs applied to a non-linear function (usually a sigmoid). The RAM neuron has been called 'weightless' (Aleksander [2]) since the output state is determined directly from one state selected from a set of stored conditions, by the input vector.

The MCP neuron cannot directly implement non-linear functions and requires additional (hidden) layers to do this. The RAM can implement any arbitrary function, but does not generalise unless additional units are provided.

The pRAM is neuron-like in that it has similar binary input and output features to the models above (Figure 2). Instead of storing the state of the output for each input vector as in the simple RAM neuron, a real number is stored which represents the probability of the neuron firing for that input vector. The output is therefore probabilistic and not deterministic. This architecture is also suitable as a hardware device as may be seen from Figure 2.

The pRAM [3,4], however can implement non-linear functions and can generalise through training in noise. The spike-train output (Figure 2) is biologically realistic as is the noise which is present at the synaptic level and not merely superimposed on the output as with some other models. These biologically-realistic features are expected to become increasingly advantageous in the applications of neural networks in respect of functionality and learning performance.

## 2  Functionality

The pRAM offers more functionality per neuron than other models. The importance of this statement is clear when hardware-realisable neural networks, comprising many neurons, are designed. If multiple layers are required to achieve non-linearity or generalisation properties, then this increases the system complexity in two areas: a) extra hardware is required to support additional neurons and b) additional connections must be provided between neurons. Although the pRAM model is more complex than the two other models shown above, the benefits of fewer layers and possibly fewer neurons to solve similar tasks outweigh the disadvantage of larger pRAM neurons. The main saving in using the pRAM is in connectivity, where the number of connections required to implement typical neural networks is large and becomes a limiting factor in the design of such networks. pRAM nets are expected to require fewer layers and fewer neurons than other neuron models to solve similar problems. Therefore savings in both silicon area and interconnections can be achieved. Other ways of reducing the connection routing problem are discussed later.

## 3  Extensions of the pRAM model

The simple pRAM model [5] described above can be extended to show other biologically plausible features which have been found to be beneficial. One extension to the pRAM model allows it to handle real numbers as input or output instead of binary numbers. In order to achieve this with a real-valued input, a spike train must be generated which has a mean firing frequency ($\in [0,1]$) over a given period equal to the real-valued number. This is naturally performed by a pRAM. If the real-valued input is stored in pRAM memory (as an 8-bit or 16-bit number) in an input layer of pRAMs for example, then if the memory location is question is permanently selected for a number of time steps, the pRAM output is defined to be 1 with a probability given by the stored memory contents. The resulting spike train is stochastic and with multiple real-valued inputs being converted in this way the input state space is well covered even though a subset of the input is used as a training set.

The internal representation of state variables is therefore in the form of spike trains. These spike trains can interact at the single spike level which leads to maximum noise in the system, or they may interact periodically when the individual spike trains are integrated over the period and therefore behave as real-numbers. The integration may be performed over a variable number of time steps to adjust the amount of noise in the system.

To convert spike trains back into real numbers at the output, the spike train is integrated over a period and becomes a real number in the range [0,1] (Figure 3).

Another extension gives the pRAM a temporal response whereby it contains a memory of recent events. This memory trace is allowed to decay with time. This is perhaps the most biologically realistic pRAM model so far. The synaptic noise is implemented by having a pRAM at each synapse and the pRAM memory can be modified so that learning may take place (Figure 4). The resultant spike trains from the synapses within the neuron are summed together with a time-constant applied to the memory element. When the output from the memory element is thresholded, a spike-train output results and is subject to a refractory period.

It can be seen that both these features are realisable in hardware with a modest increase in complexity over the original pRAM design.

## 4   Learning

The major thrust of our research has been not only to develop hardware-realisable neurons, but to develop models which allow learning processes to be performed in hardware also. A number of learning algorithms have been proposed for the pRAM. The first designs were for reward/penalty learning algorithms; and a local reward/penalty learning architecture is described here. Further research is developing back-propagation learning for the pRAM.

The local learning rule used is:

$$\Delta \alpha_{\underline{u}} = \rho((a - \alpha_{\underline{u}})r + \lambda(\bar{a} - \alpha_{\underline{u}})p)(t) \times \delta_{\underline{u},i}$$

where $\alpha_{\underline{u}}$ is the contents of the memory location addressed by the input vector $\underline{u}$, $\rho$ and $\lambda$ are the reward and penalty rates and r and p are the reward and penalty signals which in this case are generated locally and stochastically by two auxiliary pRAMs.

Figure 5 shows how the equation above is implemented in hardware. The most expensive part of the design is the multiplier, but the use of a multiplier allows learning rules to be used which are naturally bounded (ie. maintain memory contents within bounds) and which behave well for small increments or decrements of the neuron 'weight'. A design which only allows simple increments or decrements by one bit in the neuron weights does not have these benefits.

The inherent noise in the pRAM is useful in learning in two ways. Firstly, it is assumed that an input layer of pRAMs is inserted whether real or binary inputs are to be used, so that a spike-train representation exists for all signals. In this way, the test vectors have a probability distribution applied which allows the input state space to be more completely used. This is especially important for RAM-based neurons whose outputs would otherwise be undefined for inputs for which they had not been trained. Thus when inputs are applied which are not in the test set, the pRAM is likely to respond in a favourable way and will exhibit the property of generalisation. This is similar to training in the presence of noise for other neuron models, but in the pRAM the noise is inherent and does not need to be artificially created. Results of such training have shown a pRAM net to be capable of classifying patterns correctly even though these patterns have been subject to noise which may be as much as 45% of full-scale.

The second way in which noise has been shown to be beneficial is in gradient-descent training. Noise has been observed by many researchers to be a means of overcoming local minima and this is also true for the pRAM. Since the pRAM can generate a variable amount of noise, by integrating the spike trains over different periods, it has been found that conditions can be detected which lead to a slowing down of the learning process. In other words, when the derivative of the error tends to zero in any part of the network, then the pRAM noise is increased for neurons in that locality. This has been found to significantly increase the learning rate and therefore to cut the training time required.

# 5   Increasing the network size

As network size increases, all neural networks have to face the problems of scalability. This is a two-fold problem; the increase in the number of connections and an increase in required neuron area. For fully-connected feed-forward nets, if there are M neurons in one layer and N in the next, there will be $M \times N$ connections between the two layers. Linear scaling is considered acceptable. For many neurons, the cost of the addition of extra inputs increases at a less than linear rate. For RAM-based neurons however, the cost of extra inputs increases exponentially.

For these reasons, the number of inputs per pRAM neuron is usually limited to between 4 and 8. If wider inputs are required, and they usually are, then a pyramid of pRAMs is built to give the required input layer size. In this case, full connectivity can never be achieved but connection complexity is reduced as comparatively local connections only are required.

Another way of improving the scalability of the architecture in respect of connections is to employ 'virtual connections'. In this way, the number of physical connections required is small, sufficient for one set of data, but this 'bus' is used for distributing the states of neurons within the network. Figure 6 shows how this has been incorporated into a modular pRAM design. In the modular pRAM design, one custom integrated circuit and one external RAM device comprise a module containing (typically) 256 pRAM neurons. The pseudo-random number generator and comparator of the pRAM can be seen together with the learning block which performs the reward/penalty algorithm (Figure 5). Within each module, an output list contains the current state of all pRAMs in the module. Virtual connections can be made to any pRAM in the module, by selecting the appropriate pRAM output state from the output list. In order to expand the network, connections must be made to pRAMs outside the current module. This is achieved by the serial links whereby the output lists from adjacent modules are exchanged. Copies of these output lists are contained on-chip so that virtual connections may be made even to pRAMs in other modules.

A network of pRAMs can therefore be built with full availability between up to 5 modules (1280 neurons) or with local availability between an unlimited number of modules. These two configurations are shown in Figure 7.

The features of the modular pRAM may be summarised as follows, learning is incorporated in hardware, the equivalent of 16-bit digital accuracy is available for learning, 256 neurons are contained in each module at present, the net may be expanded by using the 4 serial links, a cycle time of less than $5\mu$seconds is obtained and an economical VLSI implementation is achieved by the use of external RAM.

144

# 6  Conclusion

The pRAM has been shown to be a biologically-realistic model of a neuron. It is capable of extension without losing its hardware-realisable features. Uses of both the non-linearity and stochasticity have been demonstrated. Because of the increased functionality in the pRAM neuron, larger and more economical networks can be built when compared to other solutions.

# 7  References

[1] W S McCulloch and W Pitts, "A logical calculus of the ideas immanent in nervous activity", Bulletin of Mathematical Biophysics, 5, 115-133, 1943.

[2] I Aleksander, "Microcircuit learning computers", Mills and Boon, London, 1971.

[3] T G Clarkson, D Gorse and J G Taylor, "Hardware realisable models of neural processing", Proc. 1st IEE Int. Conf. on Artificial Neural Networks, London, 242-246, 1989.

[4] T G Clarkson, D Gorse and J G Taylor, "From wetware to hardware: reverse engineering using probabilistic RAMs", Journal of Intelligent Systems, Freund, London, in press.

[5] T G Clarkson, D Gorse and J G Taylor, "Biologically plausible learning in hardware realisable nets", Proc. ICANN91 Conf., Helsinki, 195-199, 1991.

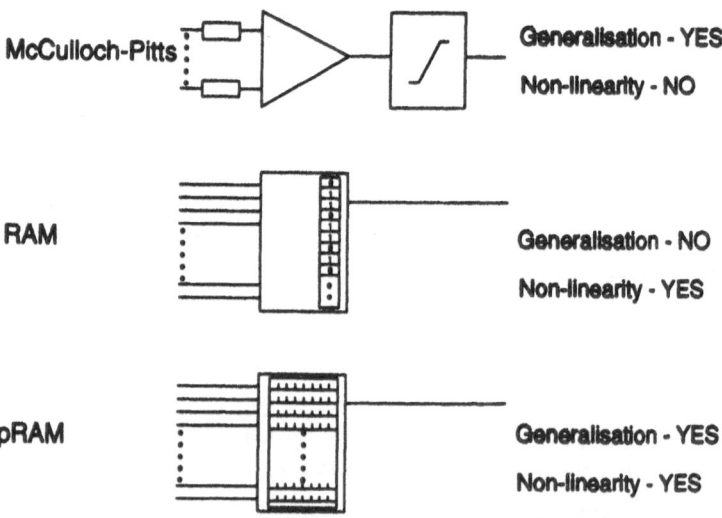

Figure 1: a) McCulloch-Pitts, b) RAM-based neurons, c) pRAM neurons

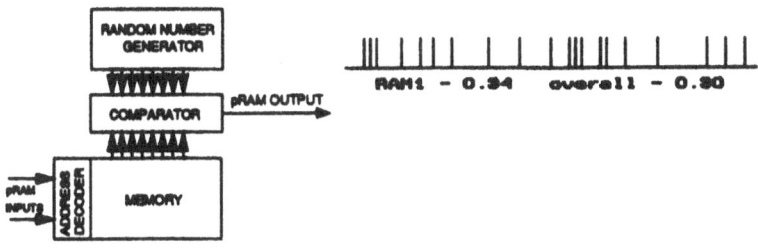

Figure 2: pRAM architecture and spike-train output

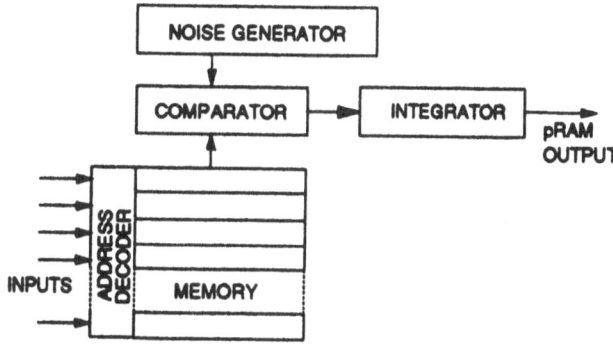

Figure 3: The integrating pRAM (i-pRAM)

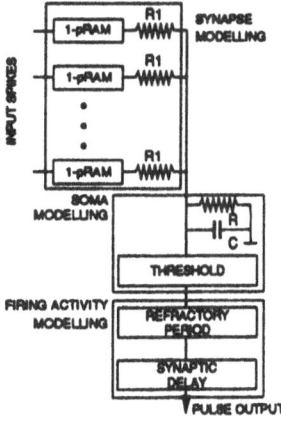

Figure 4: The noisy-leaky integrator neuron

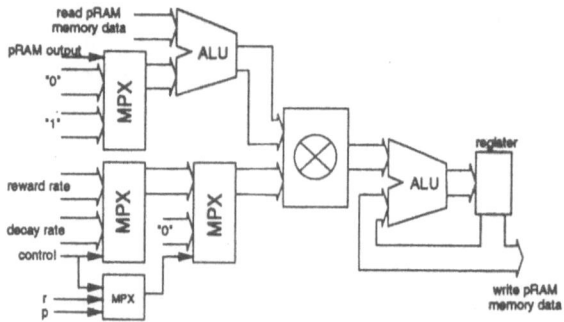

Figure 5: The pRAM reward/penalty learning block

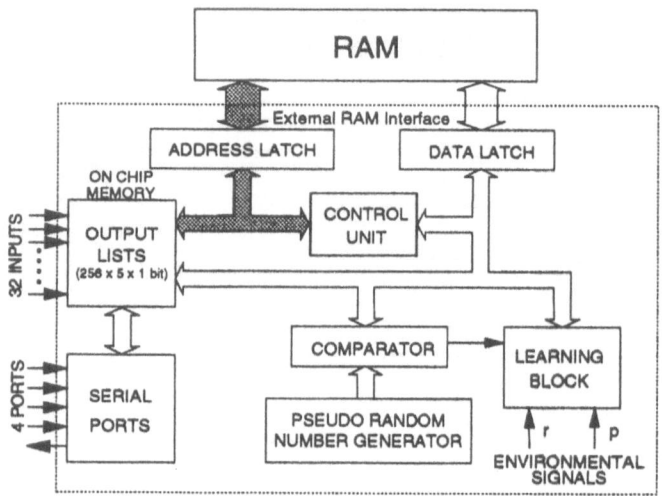

Figure 6: pRAM modular architecture

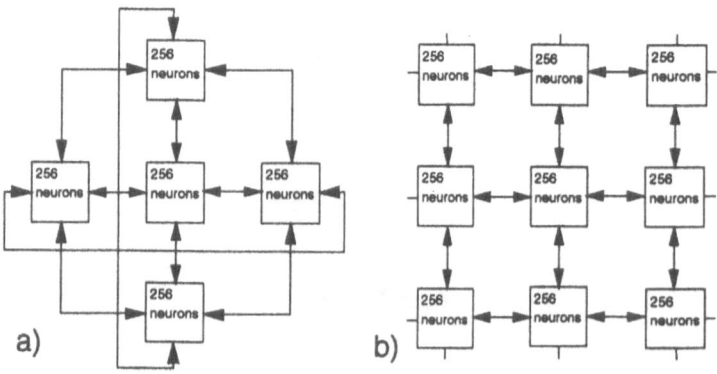

Figure 7: a) 5 module cluster with full availability and b) unlimited expansion

# The Cognitive Modalities ("CM") System of Knowledge Representation - the "DNA" of Neural Networks?

Ronald E. Wright

Wright Associates
4 Trenton Close
Frimley, Surrey
U.K.

## Abstract

A great deal of research effort is being directed towards the processing involved in Neural Networks, but relatively little attention is being given to the way information is represented in Neural Networks. This paper outlines a methodology for encoding the cognitive and syntatic bases of all natural languages, referred to as the Cognitive Modalities ("CM") system of knowledge representation. The application of this system in the automatic translation of text and in information retrieval is described. It is suggested that a similiar system of knowledge representation must be used in the brain and that such a representation has analogies with the DNA molecule and that processes similiar to those of genetic evolution may be involved in creative intelligence. This leads to an attempt to define what is meant by an intelligent system, and how such intelligence might be measured.

The paper is based on the text of a talk given at the NCM'91 meeting which, being the last of the day, was intended to be entertaining, easy to assimilate, and thought provoking.

## 1    Introduction

Just over a year ago the author was made redundant and forced into early retirement. He found this a somewhat traumatic experience and ended up in the psychiatric ward of a modern hospital. This allowed him to observe current methods for the treatment of mental illness from a "user's" stand-point. He himself really only needed a period of support and TLC (Tender Loving Care) while he got his bearings. However the Senior Consultant was keen to apply other methods of treatment and our author had some difficulty in resisting his efforts. Most patients were subjected to drug therapy, and a range of drugs were applied which went under misleading and deceptive titles, such as "tranquillizers" and "anti-depressants". Their use is rather like trying to cure a fault in your PC (Personal Computer) by spraying penetrating oil through a suitable apperture; it might cure the fault, but the side effects are difficult to predict. The author was prescribed a drug which is the subject of litigation in the United States on the grounds that it can induce suicidal and homicidal tendencies. As one of its users gunned down several people in a Canadian super-market and then took his own life, there would appear to be some substance in the allegation.

Another favoured method is ECT (Electro-Convulsive Therapy). This is rather like trying to cure your PC by connecting two wires to it somewhere and then applying the electric mains. It might cure the fault, but your PC is unlikely to ever be the same again!

A fourth treatment, Lobotomy, is akin to curing your PC fault by cutting one of the main internal bus cables. It may cure the fault, but the configuration and performance of your PC will be permanently altered.

One can conclude that the tools the medical profession has available to cure mental illness are pretty blunt instruments and they are really at a leeches and ducking-stool stage. But although one can criticize the medics' understanding of the brain, they and the other disciplines which are studying the brain have produced enough books and papers on the subject to more than fill a good sized room. However, if we ask ourselves the question: "Do we understand how the brain works?", the answer must be: "Not really". But we do have a few clues, and here are three of them.

One is the modelling of mental processes at the functional level, as instanced by the late David Marr's work on vision (reference 1). This gives us an idea of the processes the brain must achieve, but not the mechanism by which it achieves it.

Most of the other papers given at the NCM'91 meeting involve modelling at the component or neuron level. However such models are usually homogeneous and simplistic in order to make the mathematics manageable. The work of Professor Walter Freeman and others on the olfactory system (reference 2) has shown that inter-connections in the brain are complex and its processing strangely chaotic. Further development of models of individual neurons and clusters of neurons should give us a better insight into the mechanisms of the brain.

The third area cited here is knowledge representation. Much work has been directed towards processing by Neural Networks, but relatively little attention has been given to the associated data structures. Much work on knowledge representation has been done during the development of Data Base Management Systems and Expert Systems, but this has been done in the context of conventional computer memory addressing systems. The brain is different. In the brain the physical location is the address (reference 3).

The purpose of this rather personal introduction is to emphasize to workers in the field of Neural Networks that the pressing need is not for the invention of clever systems and gadgets, useful though these may be, but for a better understanding of the nervous system on which valid medical treatments can be based.

## 2    The Cognitive Modalities System

Some five years ago the author made the acquaintance of Professor Douglas A. Young who was working in the Department of Computer Science of the University of Manitoba. He had pondered on the lack of success in the development of automated language translation. Despite the application of considerable research effort and computer processing power to the problem, proper language translators are still not commercially available.

Douglas Young took the view that most of our knowledge is acquired through our senses and actions, and there is evidence that we ourselves, and certainly our domestic pets, do not handle all our knowledge in the form of words. In 1983 he came up with a theory for a cognitive, non-verbal representation of information. This asserts that the meanings of words and sentences can be represented in terms of weighted aspects of the fundamental categories of cognitive information, which he refers to as the Cognitive Modalities (CM), modality refering to one or more of the different categories of information that the brain processes (see references 4 and 5). He and his team in Winnipeg developed this idea to provide a system which could represent every word in the English language modally in all its different meanings and uses. This CM system uses a cognitive lexicon of some 8,000 modality specific elements or codes grouped according to the sub-modal categories of perception and non-sensorimotor events.

Typically the semantic and syntactic features sufficient to define the meaning of any given word can be represented by some ten modal codes (each of three or six characters). Two 600 word English to CM dictionaries were produced in an earlier project. The modal subcodes have now been completely revised and improved, however, and a new dictionary is being produced, utilising an X-Window based machine-assisted system of modal coding. It is possible in principle to produce dictionaries for other languages to CM code, so that the CM code offers a universal language, or "interlingua", for machine translation.

Douglas Young set up a small company in Winnipeg to develop machine translation and this produced a demonstrator which could translate a specific sample of legal text from English into French via the CM code at an average speed of about one word every 15 milliseconds, including on-line disambiguation, with a reasonable accuracy (typically half of the text produced did not require editing). The programmes were written in Zeta Lisp and ran on a Lambda Lisp machine.

An important feature of the CM representation of a word is that its contents are meaningful and can therefore be content addressed and used directly as a basis for reasoning. For example the code for "dog" might include a code element specifying the colour of its hair, so it would be possible to content address for white dogs and reason about them. The system at Winnipeg was able to give limited demonstrations of such capabilities.

The CM system is comprised of four principal segments. The first is a cognitive lexicon, COLEX, which identifies and codes more than 8,000 modality specific elements of cognitive information grouped according to the appropriate modality. It is based upon the 32 principal modalities listed in Table 1 (appended). These modalities are based on categories of cognition, such as touch, or on the mental experience of information, such as time. The Metaconceptual modality forms the largest category of major sub-modalities, grouped under the headings of Classification Fields, Syntactic Indicators, and Semantic Indicators.

The second segment of the system is a modality-specific English dictionary called MODIC, which incorporates both semantic and syntactic information against each word listed. Table 2 gives the format of the CM meaning profile of a given word and it comprises 22 different elements, several of which are optional. Each word is classified in three basic ways. Firstly according to its principal cognitive feature, called its "use-category". Secondly according to its principal grammatical feature, referred to as its "use-category form", which will be one of six sub-groups of conventional grammatical types. Thirdly according to its "meaning classification field", which cites the word as occuring in a particular location within a comprehensive set of tree structures that relate any given word and each of its meanings to its superordinate and subordinate terms. For example the word "furniture" would have "artificial objects" as superordinate and "chair" as a subordinate. Each element in Table 2 has sub-codes, for example "cognitive category" has some 40 sub-codes.

The third segment of the CM system is an inter-active disambiguation program called MIDA. This can perform on-line disambiguation. Just as we ourselves, when trying to understand a sentence, use a relatively small amount of the knowledge available to us, so MIDA uses only a limited amount of the modal information available in performing disambiguation.

The fourth segment of the CM system, OPEL, is in development. It will represent each word-form in orthographic (or written) terms and phonological (or spoken) terms.

## 3    Current Status of the CM System

Douglas Young is currently working with Unitran Systems Inc in Ottawa on a contract funded by the Canadian Department of Defence aimed at building a demonstrator information retrieval system based on the CM system. The demonstrator is being designed to respond in natural language to questions about information contained in stored text and also to provide appropriate text extracts from the stored literature.

Data retrieval systems which use Neural Network techniques for the retrieval of text (such as Excaliber) and relational data base machines (such as the Teradata 1012), which can handle large volumes of data, are commercially available, and the combination of these with the CM system would appear to have considerable potential.

This paper is based on the author's understanding of the CM system being developed by Dr Douglas Young and his team, and may therefore be unintentionally misleading in some respects.

## 4    The CM System as a basis for Creative Intelligence

Table 3 is an annotated example of the modality codes for a complex activity, namely "flew", as related to an air passenger (reference 6). For the purposes of the present argument you need only note that it is a cluster, set, or association of coded elements. It has struck the author that there is a loose analogy between this method of representing words and concepts and the genetic structure of the DNA in the chromosomes which defines the form of living organisms. This leads to an hypothesis which is expressed as "Wright's Fifth Law of Neural Networks", as four other "laws" have been proposed in an earlier paper (reference 7). Here it is:

"The brain stores knowledge in whole or in part in some form of cognitive modality code, so that thoughts can be represented by associations of such codes and the evolution of thought (creative thinking) can be the result of modifications and associations of thought patterns analogous to genetic mutation and evolution. This implies some form of selection or optimisation process to represent genetic "survival of the fittest" (or "survival of the stable") selection."

Well, how can this assertion be justified? Convincingly, perhaps, only by producing a satisfactory model of the process. But in the meantime there are some general comparisions between genes and Cognitive Modalities which tend to support the idea as follows:

A combination of genes defines an organism (animal, plant, etc).
A combination of CM elements defines a word, concept, thought.

Genes use a sub-code of three bases.
The CM system uses a sub-code of three or six alphabetic characters to define some 8,000 modality specific elements.

Some hundreds of thousands of genes define an organism.
Nouns and verbs use up to a few hundred CM elements per meaning.

The genes of an individual are part of the gene pool of the population.
The knowledge of a person is part of the total knowledge of the population.

A characteristic of an individual (e.g. colour) might be determined by a specific gene.
A characteristic of an individual (e.g. colour) might be described by a single modal code element.

(N.B.: This is a very significant analogy as there can be a one-to-one correspondence between the two systems, a change of a genetic code requiring a corresponding change to the CM coding.)

Evolution of an organism can be effected by the change of one or more genes in the DNA molecule (see reference 8).
Modification of thoughts might be effected by the change of one or more of the cluster of modal codes defining the thought.

Competition between organism variants implies a selection process.
Competition between thoughts, plans, desires, objectives implies a selection/optimisation process.

We do not fully understand the genetic process (e.g. dominant genes).
We do not fully understand the mechanisms of creative thought (e.g. musical composition).

## 5    Intelligence and Intelligence Metrics

In the above the author has been trying to define a mechanism for intelligence without first defining what was meant by intelligence, and this is an almost universal failing of books and papers on artificial intelligence. The Concise Oxford Dictionary gives it as "intellect, understanding".    Aggarwal and Duda imply that intelligence comprises perception, reasoning, learning and understanding (reference 9). A definition of an intelligent system might be that it is a system that is capable of modifying its enviroment to its needs, rather than just adapting to its enviroment as does a plant. In living things we can associate intelligence with organisms which have neurons, which leads the author to tentitively propose his "Sixth Law of Neural Networks", namely:

"An "intelligent system" is one which can or does modify its enviroment to its needs. In living organisms such intelligent behaviour requires neurons; in non-living systems such intelligence requires a capability for self-organisation and learning as may be provided by an artificial Neural Network."

The author has discussed the intelligence of systems and how such intelligence might be measured in an earlier paper (reference 10). Here he would just suggest to you, the reader, that next time you use the word "intelligence" to describe an inanimate object you should consider how the intelligence manifests itself and how it can be demonstrated or measured.

## 6    Conclusions

There is no doubt that knowledge representation must be considered in the study, emulation, and realisation of systems involving Neural Networks. The CM system, although still in development, would appear to be based on firm foundations and may well be a key to the effective realisation of classes of system which have proved difficult to achieve using conventional computer and A.I. technology. Such candidate systems include the machine translation and natural language retrieval systems discussed in this paper as well as other systems for other applications such as sensor signal description, situation assessment, decision support, and automated surgery.

Knowledge representation certainly warrants more attention than it is currently getting from academia, government research laboratories and industry in the U.K. Moreover the study of knowledge representation in a cognitive context should progress our understanding of the human brain and provide a sounder scientific basis for the treatment of mental health than the pseudo-science engendered by Sigmund Freud and others on which our current treatments are based.    This pseudo-science is being discredited, but there is as yet no established and comprehensive body of brain science to replace it, and this poses a major challenge to the medical and scientific research communities.

## 7    Acknowledgements

The author's thanks are due to the late Mr Basil Z. de Ferranti for encouraging the investigation of Neural Networks at a time when being "First into the Future" was no longer fashionable, Mrs Betty Rieden for her briefing on genetics and intelligence, and Mrs Sheila Oelman for providing some TLC when he badly needed it.

## 8    References

(1)    D. Marr, "Vision", Freeman, Cooper and Co., San Francisco, 1982.

(2)    Walter J. Freeman, "The Physiology of Perception", Scientific American, February 1991, pp 34-40.

(3)    T. Kohonen, "Self-Organisation and Associative Memory", Third Edition, Springer-Verlag, 1989, ISBN 0-387-51387-6.

(4)    D. A. Young, "Language and the Brain: Implications from New Computer Models", Medical Hypotheses, 9:55, 1982.

(5)    D. A. Young, "Interactive Modal Meaning as the Mental Basis of Syntax", Medical Hypotheses, 10:5, 1983.

(6)    D. A. Young, private communication.

(7)    R. E. Wright, "The Potential Applications of Neural Networks and Neurocomputers in C3-I", Science of Command and Control: Part II, AFCEA International Press, 1989, ISBN 0-916159-18-3, pp 159-164.

(8)    Richard Dawkins, "The Selfish Gene", Oxford University Press, 1989.

(9)    J. K. Aggarwal and R. O. Duda, "Guest Editors' Introduction", IEEE Expert, Vol 1, No 1, Spring 1986, pp 10-11.

(10)    R. E. Wright and D. A. Young, "The Cognitive Modalities ('CM') System of Knowledge Representation and Its Potential Application in C3I Systems", Proceedings of the 1990 Symposium on Command and Control Research, SAIC Report SAIC-90/1508, pp 373-381.

**Table 1: Sensory and Mental Modalities**

Primary Sensorimotor Modalities and Sub-Modalities

1. (VPA) Visual pattern; 2. (VDM) Visual detection of movement; 3. (COL) Colour; 4. (RIL) Retinal illumination; 5. (VRA) Visual ranging and depth perception; 6. (VTD) Visual tracking and direction sensing; 7. (AUD) Auditory pattern; 8. (ADS) Auditory direction sensing; 9. (PCP) Proprioception; 10. (TAC) Tactile; 11. (NCP) Nociception; 12. (TMP) Temperature; 13. (OLF) Olfactory; 14. (TST) Taste; 15. (MOT) Motor.

Compound Sensorimotor Modalities

16. (HAP) Haptic; 17. (GUS) Gustatory; 18. (ACP) Autonomic control and proprioconception; 19. (VES) Vestibular; 20. (STE) Stereognostic; 21. (CMD) Command.

Mental Modalities

22. (MET) Metaconceptual; 23. (TIM) Mental time; 24. (ISP) Introspection; 25. (EMS) Emotive mental states 26. (CMS) Cognitive mental states; 27. (CMA) Cognitive mental states; 27. (CMA) Cognitive mental acts.

Compound Sensorimotor/Mental Modalities

28. (PLX) Phonological expression; 29. (PLP) Phonological perception; 30. (OGX) Orthographic expression; 31. (OGP) Orthographic perception; 32. (MMR) Multimodal memory retrieval.

**Table 2: Compostion of One Word Meaning in Modal Dictionary**

Schema (XKS)

YYC Cognitive category (1 submodality code).
YMD Description (2 to 8 subcodes, describing entity, activity, characteristic, metaconcept).
YMF Function: how it does what it does (typically 2 to 6 subcodes).
YMP Function: Purpose, goal, or use (typically 2 to 6 subcodes).
YMO Function: Operation, modus operandi (typically 2 to 6 subcodes).
YMR Function: Requirements, essential needs (typically 2 to 6 subcodes).
YMC Function: Constituents, parts, or elements (typically 2 to 6 subcodes).
XXF Classification field (1 to 6 subcodes).
YMS: Context (1 to 8 subcodes).

Syntatic (XKY)

YVV Cognitive valency: expectation or need for proceeding or following YYC (1 to 4 subcodes).
YYF Use-category form (1 subcode).
TIM Tense (2 subcodes).
YU- Voice (1 subcode).
YM- Mood (1 subcode).
YI- Case (1 subcode).
YQ- Number and/or person (1 subcode).
YK- Gender.
YM- Semantic indicator: excluding those under Schema (1 subcode).

Lexical (XKL)

PLX Phonological expression.
PLP Phonological perception.
OGX Orthographic expression.
OGP Orthographic perception.

**Table 3: CM Meaning Profile of Air Passenger Meaning of 'Flew'**

YYC: yah  (Cognitive category:# complex/metaconceptual).

YMD: yup[xmf.mcn] (Description: # Be conveyed (by machine)).
  ste.dth[xoi.mar]  (# Through the air).
  cms.exp[ves,vso,vso];  (#(experiencing)  vestibular  changes  of  speed,
pressure,orientation).
  [tac.ppv[rtp,rtb]]  (# Variations in tactile pressures on seat or back).
  [aud.xft];  (# In aircraft noises).
  [vpa.vai]  (# In-aircraft sights).

YMF: cms.jcd.nev[xmf.dfp[xoi.mar]] (Function:# differential of air pressure).
  cms.nev.sco[xen.egn];  ((how) # opposing gravity).
  ymm[xmf.def[xop.mar];  (# (caused by) machine-induced airflow).
  ste.dar[xoa.mtp.asa]  (# around aerofoil).

YMP: cms.nev.bap  (Purpose: # Passage).
  [ste.pdl;  (# Long-distance).
  tim.rvf]  (# High-speed).

YMO: mms.ufl  (Mod. operandi: # proceedures of flying as a commercial flight
passenger; a modal version of a Shankian script).

XXF: xcv.taa  (Classification: # activity, travel, air (unspecified vehicle type).

YMS: xev.hdd,hwd,hwb,heg,hdh,hdv (Context: # domestic, political, business, emergency,
hobby).

[The syntactic part (XKY) of the word-meaning profile comprises:]

YVV: yva.yet,yel,ydd,ydr,ydt,yc*  (Cognitive valancy: # as not next word must be
followed, next or near, by a temporal or abstract entity-word, a determiner or
substitutional descriptor, or any of a wide range of connectives).

YYF: (Use-category form: # intransitive verb).

TIM: (Tense: # past simple).

# AUTHOR INDEX